PLANT DISEASE CONTROL

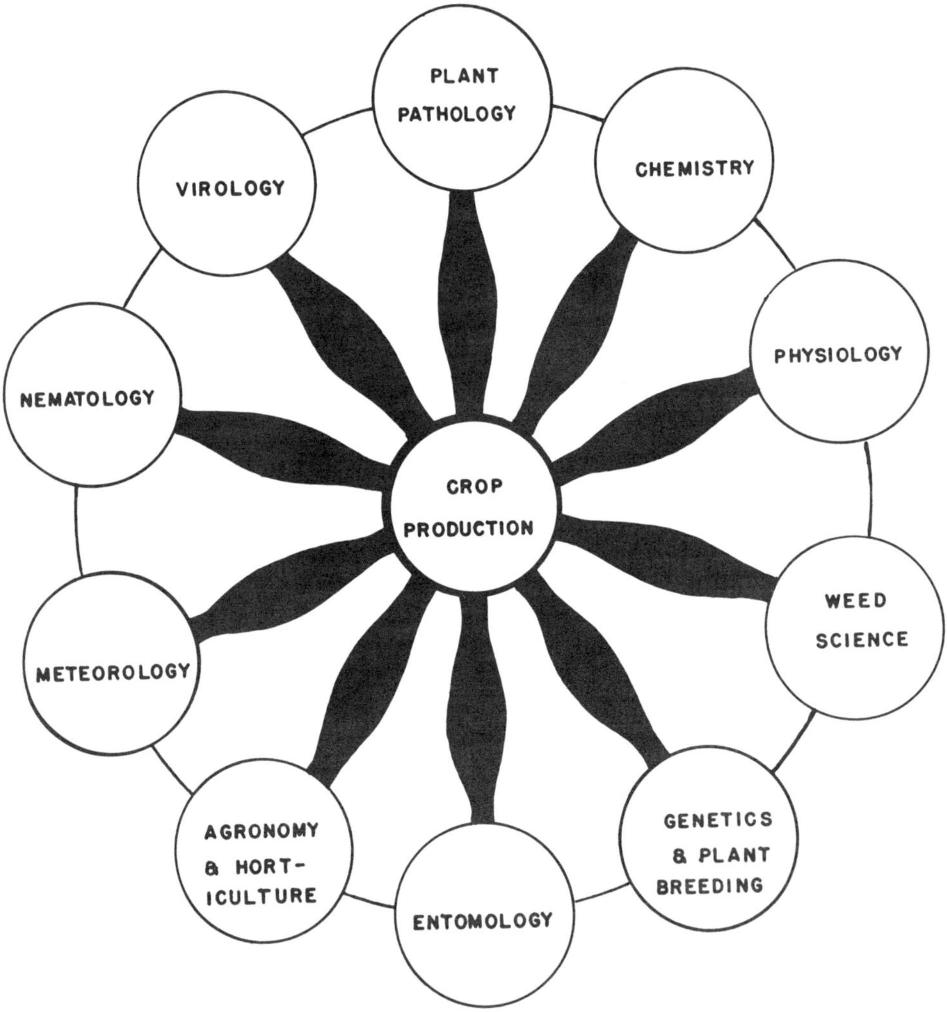
PLANT PATHOLOGY
CHEMISTRY
PHYSIOLOGY
WEED SCIENCE
GENETICS & PLANT BREEDING
ENTOMOLOGY
AGRONOMY & HORT-ICULTURE
METEOROLOGY
NEMATOLOGY
VIROLOGY
CROP PRODUCTION

PLANT DISEASE CONTROL

Eric G. Sharvelle, Ph.D.

Professor Emeritus
Purdue University
Lafayette, Indiana

avi

AVI PUBLISHING COMPANY, INC.
Westport, Connecticut

Library of Congress Cataloging in Publication Data

Sharvelle, Eric George.
Plant disease control.

Includes index.
1. Fungous diseases of plants. 2. Fungi, Phytopathogenic—Control. 3. Fungicides. I. Title.
[DNLM: 1. Fungicides, Industrial. SB951.3 S532p]
SB733.S46 632 79-16417
ISBN 0-87055-335-6

Printed in the United States of America by Eastern Graphics, Inc., Old Saybrook, Connecticut

Preface

In this world of energy crises, oil spillage, nuclear proliferation, computerized technology, social change, and continuing international controversy, it is difficult to realize that man is still a guest of the plant kingdom on this orbiting satellite known as EARTH.

Many representatives of *Homo sapiens*, either in ignorance or in overfed complacency, take flowering plants for granted and have little concern for, or understanding of, their significance for the welfare of mankind. Even though astronauts from different nations can ride their juggernauts of outer space to meet and exchange greetings hundreds of miles above the surface of the earth, most common people must eat to survive. DAVID FAIRCHILD, the renowned American explorer, once declared, "Agriculture is the source of that element which Plato declared was the most important of them all in the building of a state—FOOD."

Despite our vaunted technological progress in industry and agriculture, the provision of sufficient food for ourselves and for others in the immediate future is a cause for grave concern for agricultural scientists and for the governments of developing nations with exploding populations. The future provision of our daily bread will be no easy task and few people appreciate the technology that is required to provide the thousands of processed food items displayed in the average supermarket. There are many factors that may and do limit world food production. Among these, the "Crop Robbers," plant diseases, must be considered as important saboteurs of essential crop yields.

For the commercial grower of wheat, rice, soybeans, maize (corn), and fruits, and vegetables—the farmer—a crop failure or a significant reduction in productivity may mean the difference between profit and loss. Many farmers in the major corn-producing regions of the world still remember the losses caused by southern corn leaf blight (*Helminthosporium maydis*) in 1970-1971 that devastated fields of corn in the U.S. Corn Belt and other regions. Many people who once admired the stately beauty of the native American elm have mourned its virtual destruction by the Dutch elm disease, and regret the almost complete elimination in the U.S.A. of chestnut trees as a consequence of chestnut blight disease (*Endothia parasitica*). These are well known examples of the destructive potential of plant diseases that may attack and destroy many plants both large and small.

In agriculturally developed countries crop losses caused by plant diseases may be tolerated in times of plenty, but in less well developed countries, or following years of low crop yields, these losses become a source of concern for those involved in the task of providing adequate food supplies for the six billion people who will populate the earth at the dawn of the 21st Century.

For the grower of plants for pleasure, it is always a source of frustration and disillusionment to discover, with experience, that most plants in the home garden rarely look like they are pictured in the seed catalog, and often fall victim to numerous pests and adverse growing conditions. Only infrequently will garden plants grow to maturity without suffering from environmental hazards that conspire to mar their beauty, reduce their productivity, and sometimes even threaten their existence. When this happens, the Green Thumb Enthusiast is puzzled and concerned by the parade of ragged roses, puny petunias, tattered tulips, tired turf, spotted strawberries, scabby apples, and other problems that frequently afflict the dwellers of the backyard garden.

To prevent plant diseases effectively, a knowledge of their nature and cause is fundamental to successful control. Once the causal agents have been correctly diagnosed, then it is possible to prescribe effective methods for their prevention. However, there are no miracle cures for most plant diseases, and, in fact, most plant diseases cannot be cured once they are well established. However, they usually may be prevented by a combination of plant protection principles that are usually necessary for effective protection against the several agencies that may incite plant diseases.

Chemicals for plant disease control, the fungicides, are just one of the weapons in our arsenal of defense against unwanted and frequently destructive blights of growing plants. Only infrequently will fungicides work alone to prevent plant disease outbreaks, but when used in combination with other methods they are safe and effective remedies for many plant ills.

In recent years, usage of chemicals in agriculture, and in the home garden, has become subject to a veritable barrage of adverse criticism by organic gardeners, some governmental agencies, radio and television performers, a few politicians, and publicity-seeking journalists. Agricultural chemicals, including fungicides, have become victims of stringent federal and state regulations under the provisions of the Federal Insecticide, Fungicide and Rodenticide Act (FIFRA). Very recently, still further restrictions have been placed on the manufacture and distribution of agricultural pesticides under the authority of a process known as "Rebuttable Presumption Against Registration" (RPAR) which requires a detailed assessment of risks and benefits of all pesticides for which there is some evidence that they may cause unreasonable risk to man or the environment. There are now very few commercial products subject to the restrictive regulations governing agricultural pesticides, including fungicides.

Fungicides, chemicals for plant disease prevention, are included by a large segment of the public in the same category as those insecticides known to be toxic to humans. Fungicides have been branded along with insecticides by an uninformed public as "Poisons on the American Dining Table." The controversy on the desirability of using chemicals on essential crop plants will doubtless continue and this textbook will inevitably become a target for criticism by those opposed to the use of any chemical on crop plants that are grown for human or animal consumption.

Ever since the publication of the late Rachel Carson's "Silent Spring" in 1962, there has been a public plethora of fiction and a paucity of facts printed on the nature and safe uses of chemicals for plant disease control. In the minds of many of the laypublic the word "fungicide" is either synonymous with "insecticide" or they have never heard of the word. In fact, fungicides, the plant disease protectants, are usually to be found in the "Insecticide Department" of most department and garden supply stores.

Despite the unfavorable publicity and lack of general knowledge concerning the necessity for plant disease control chemicals, professional fruit and vegetable growers, the commercial florists, and the informed home gardener have accepted and successfully used fungicides as useful tools in the annual Battle of the Blights on the farm and in the suburban garden.

As early as 1939 the author was working with the distinguished fungicide expert, Dr. James G. Horsfall, at the Geneva, New York and the New Haven, Connecticut experiment stations in the search for safer and more effective organic chemicals for plant disease control. As a result of this research the first organic fungicide, a derivative of quinone named "Spergon," was introduced by the U.S. Rubber Company (now known as Uniroyal Chemical Co.). The introduction of this organic, non-mercurial seed protectant was credited with opening the door for further discovery and introduction of organic chemicals of low mammalian toxicity for the control of plant diseases. Since 1939 virtually all of the modern fungicides have been organic compounds.

In this book an attempt has been made to organize the available information on modern tools for the prevention of losses caused by plant diseases. It is hoped that this compilation of fungicide facts will help to draw aside the curtain of confusion and misinformation generated by overzealous environmentalists and misinformed conservationists that has cast a stigma of unfounded belief that chemicals for plant disease control, like some other pesticides of established mammalian toxicity, are poisons on the dining table of the American homeowner, and therefore should be barred from use by professional growers of edible plant products.

For more than a quarter of a century the author served as the "Hoosier Plant Doctor" at Purdue University with the major responsibility for preparing plant disease control recommendations for Hoosier farms and backyard gardens. This reference was prepared with the assistance of many professional friends in land-grant universities and the agricultural chemical industry. Their help and contributions are recognized in the Acknowledgements. Appreciation is expressed also to Dr. Norman W. Desrosier and Ms. Karen L. Carter of the AVI Publishing Company for their encouragement and assistance in bringing this book into being. The writer is the author of two other textbooks on plant disease control published in 1960 and 1969, and hopes that this modern reference on plant disease control will constitute a useful reference for professional agriculturalists, university and vocational agriculture teachers, county agricultural agents, representatives of the agricultural chemical industries, and for those who enjoy the pastime of growing plants for pleasure.

Eric G. Sharvelle

Lafayette, Indiana

March 1979

Acknowledgements

Three difficulties often are incurred in authorship: to write a concise, informative manuscript; to locate an effective publisher; and to encourage wise people to read the final results. With the help of those listed below, I have attempted to surmount the first two problems, and hope that the last difficulty will be resolved.

To Purdue University, I am grateful for the privilege of over thirty years' membership in its distinguished family of "people helpers."

I am also indebted to the following plant pathologists for their contributions of specific information on plant disease control:

Dr. Einar Palm, University of Missouri, Columbia, Missouri
Dr. Walter Stevenson, Purdue University, Lafayette, Indiana
Dr. Donald Scott, Purdue University, Lafayette, Indiana
Dr. Arthur McCain, University of California, Berkeley, California
Dr. David Quinn, West Virginia University, Morgantown, West Virginia.

To Dr. Donald K. Tressler and Dr. Norman W. Desrosier and their associates in the AVI Publishing Company, my sincere thanks for their acceptance of the manuscript for publication. The professional competence and patience of Ms. Karen L. Carter, AVI Publishing Company, in editing and correcting the original manuscript has made this difficult task a much improved and pleasant undertaking.

Finally, to "Glib," my partner of many years in many places, a long deferred acknowledgement for her patience, tolerance, and encouragement, without which this project could never have been completed.

Contents

1

Introduction to Plant Disease Control

In the last quarter of the 20th Century man continues his struggle for survival and dominance over all other living creatures on the earth. Unfortunately, much of man's struggles during the third quarter of the century were devoted to a contest for dominance of man over man, with resultant conflict between competitive nations. It is difficult in this age of technology to appreciate that at one time science as we know it today did not exist.

In the beginning darkness covered the face of the earth, and science developed slowly, at times painfully, through eons of time. There still remains more of the creator's undisclosed secrets in the centuries to come. As continuing scientific progress moves human society towards new horizons of exploration, scientists will continue to unscrew the inscrutable in the never-ending seeking of truth.

The history of modern science, like life itself, has consisted of a molting of our illusions and the dissolution of our delusions. If we are to consider ourselves as scientists we must continuously be prepared to form theories and creeds today, only to discard them tomorrow, as new principles and practices develop as the products of scientific investigation. We must constantly keep in mind the realization that today's tools will not always suffice for tomorrow's needs. Annually the Canadian blue goose molts its feathers because new ones have developed to replace them. Science has progressed slowly up the stairway of increasing knowledge, formulating theories, classifying and dovetailing facts, creating new principles, discarding disproven laws and axioms, and applying technology to practice in the never-ending struggle for survival in an increasingly competitive world.

Abraham Lincoln, 16th president of the U.S.A., speaking before its 1862 Congress in support of the Morrill Act proposing the establishment of the land-grant colleges, made the following statement:

No other human occupation occupies so wide a field for the profitable and agreeable combination of labor with cultivated thought as agriculture. Every blade of grass is a study; to produce two where but one grew before is both a profit and a pleasure.

The thought recurs that education—cultivated thought—can best be combined with agricultural labor, or any labor, on the principle of thorough work, and ere long the most valuable of all arts will be the art of deriving a comfortable subsistence from the smallest area of land.

With these words, in a time when farm land was plentiful and 85% of the people tilled the soil to provide sustenance for the remaining 15% of the people in the U.S.A., President Lincoln outlined for posterity the true purpose of the U.S. Land-Grant philosophy. With the passage of the Morrill Act by Congress in 1862, Justin Smith Morrill, the U.S. senator from Vermont, established the beginning of a system of agricultural education that was destined to serve as a model for agricultural development in America and many other countries of the world.

When Abraham Lincoln spoke of "deriving a comfortable subsistence from the smallest area of land," he could not have visualized the problems that would be encountered today in providing sufficient food for exploding world populations in the 20th and 21st Centuries. Neither President Lincoln nor Senator Morrill could have visualized the day when growing industrialization, mushrooming cities, expanding interstate highways, shopping centers, and other products of modern social progress would swallow up a million acres of tillable land in an era when 10% of the national population would be able to produce daily bread for the remaining 90%, with crop surpluses left over to feed some of the hungry in other agriculturally developing countries. They surely could not have foreseen the day when the U.S. Government would spend $500 million for agricultural research and education as well as providing additional moneys to support agricultural development programs in several hungry areas of the world.

Abraham Lincoln, with all of his widsom, could not have predicted the day when concentrated orange juice would be frozen and shipped thousands of miles to be consumed many days later as fresh as the day the fruit was harvested. The pioneer workers in the early days of the U.S. Department of Agriculture could not have visualized the use of hormones to manipulate plant growth, nor the use of antibiotics to develop faster feathering and earlier laying poultry, nor the use of antibiotics for the control of some plant diseases.

THE SCOPE OF AGRICULTURE

Agri-business currently is a major industry in many countries of the world. Daniel Webster, statesman, lawyer, orator, and U.S. Secretary of State in 1850-1852, once wrote, "Let us never forget that the cultivation of the earth is the most important labor of man. When tillage begins other arts follow. The farmer therefore is one of the founders of civilization."

Modern agriculture is no longer the art of animal and crops husbandry, where a knack with horses and mules, a green thumb, and an eye for the weather suffice to ensure success for the tiller of the soil producing essential food crops. Agriculture is now a computerized technical profession having the responsibility for producing sufficient food to feed the world's expanding millions, and is one of the foundation stones on which world peace and human progress must be erected and preserved.

Although many acres of land in the world are devoted to non-food production uses, it has been estimated that 80% of the world's cultivated land is devoted directly or indirectly to crop production for human and animal consumption. Although widely publicized reports indicate that farm numbers are declining, the true overall picture is missed. According to a National Agricultural Chemicals newsletter of January 23, 1976, agriculture is still the nation's largest employer with some 14-17 million workers engaged in some phase of food production. Farming alone involves some 4.4 million people.

On the consumption side, agriculture is a huge customer. Annually in the U.S.A., $8.8 billion are spent for tractors, trucks, and farm equipment. American farmers annually spend $10.1 billion for fuel, lubricants, and machinery maintenance, and use more petroleum products than any other industry. They also use 360 million pounds of rubber annually, which is sufficient to equip 7 million automobiles with tires. In addition, the U.S. farm industry uses 33 billion kilowatthours of electricity, which is more than the combined total used in the New England states, Kentucky, Maryland, and Washington, D.C. U.S. agriculture annually uses 6.5 million tons of steel for the manufacture of farm machinery, farm trucks, fencing, and building materials which provides employment for 40,000 steel workers. In addition, American farmers spend $17 billion for seed, feed, fertilizer, and lime. The American farm industry is big business and should be more widely recognized as one of our most essential industries.

There seem to be but three ways for a nation to acquire wealth; the first is by war, as the Romans did, in plundering their conquered neighbors—this is robbery; the second by commerce, which is generally cheating; the third by agriculture, the only honest way, wherein man receives a real increase of the seed thrown into the ground, in a kind of continual miracle, wrought by the hand of God in his favor, as a reward for his innocent life and his virtuous industry.

Benjamin Franklin

THE WORLD FOOD SITUATION

Much has been said, and doubtless much more remains to be said, about world food requirements in the immediate future. If plant pathology is to play a role in world food production in the next 10-15 years by limiting plant disease losses through the control of major diseases, then a brief consideration of the food production problems in the next decade will serve to focus attention on this role of the science of phytopathology.

In July of 1976 a World Food Conference was held at Iowa State University, Ames. This international conference was convened to consider immediate world food production requirements and was attended by 1,600 agriculturalists from 70 different countries.

It was the consensus of opinion of the conferees at this international conference that by the end of the next 25 years where there is now one loaf of bread there will be need for two. It was emphasized that the world population is still increasing at the rate of 2.4% per year. At the present time, there are approximately four billion people in the world. Population experts predict that by the year 2000 it will be necessary to feed 6.5 billion people, all of whom will have to eat to survive. This means that world agricultural experts will have to achieve a food production increase of 75% over present production levels just to stay even. While some amateur food production experts probably believe that U.S. agriculture is over-producing, the truth of the matter is that food reserves have decreased in recent years. World feed grain and wheat reserves dwindled from 150 million metric tons in 1975 to slightly in excess of 80 million metric tons at the start of 1976.

World Food Council officials reported at the Iowa World Food Conference that at the present time 1.5 billion people are underfed. In 1965, 460 million people suffered from severe malnutrition (Fig. 1.1). This means that approximately one out of every ten people in the world is hungry, which does not add up to an oversupply of essential food.

Despite localized progress in crop production the world as a whole so far has failed to develop a large staple food production increase. For the first time in 20 years world food production declined in 1972 by 33 million metric tons of cereal grains, which would have fed 100 million people. World food production increased in 1973, declined again in 1974, and rose again in 1975, further indicating a lack of present ability to stabilize world food output under the inevitable annual variations in climatic conditions.

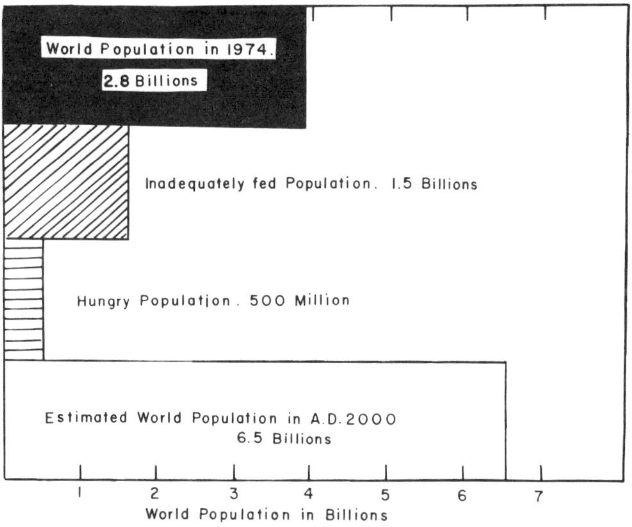

FIG. 1.1. THE WORLD FOOD SITUATION

At the present time nearly 3.5 billion acres of land, or 11% of the world's total surface area, are cultivated and used for essential crop production. It has been estimated that another 3.0 billion acres of land might be converted to crop production, but it would probably cost more than the value of the food produced.

According to Ciba-Geigy Quimica S.A., the value of world food production losses resulting from plant diseases, weeds, and insects annually amounts to $60 billion. Of this amount, $52 billion is the annual value of cereal grain losses from diseases, weeds, and insects, with a loss of $8 billion to the world potato crop (Fig. 1.2).

Ciba-Geigy Quimica S.A. reported the following crop losses in Japan from plant diseases, weeds, and insects, expressed as losses of millions of metric tons:

Wheat	86
Rye	6
Barley	129
Oats	16
Maize	121

Milo	46
Potatoes	129
Sugar beets	69
Sugar cane	566
Vegetables	78
Other Fruits	21
Grapes	28
Coffee	2.6
Cotton	5

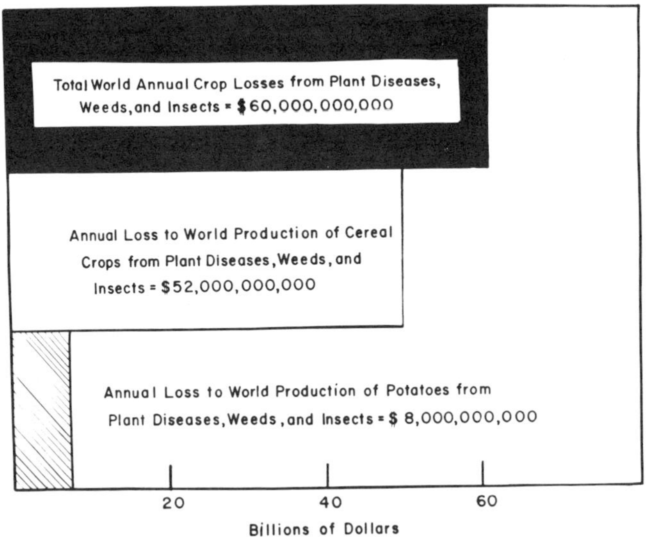

FIG. 1.2. LOSSES TO WORLD FOOD PRODUCTION CAUSED BY PLANT DISEASES, WEEDS, AND INSECTS

At the conclusion of the World Food Conference at Ames, Iowa, in 1976, Dr. Sartaj Aziz, Deputy Executive Director of the World Food Council, stated:

> The world food problem is essentially manageable, but it will take tremendous national and international effort and is not a problem of simple alternatives. The principal responsibility for preventing future food disasters rests with the United States of America.

It is readily apparent that the task of providing sufficient food for world populations in the 21st Century will be an important undertaking for those agricultural scientists dedicated to helping themselves and others in the age-old struggle of producing food for survival.

Prevention of essential crops losses caused by plant diseases will be one of the spokes in the Wheel of Food Production for the 6 billion people who will come to dinner in the year 2000. In this task, chemicals for plant disease control, the fungicides, will play an important role as partners with other practices in the food production task that lies ahead in the 21st Century.

WHAT IS PLANT PATHOLOGY

An understanding of the nature of plants is necessary for any consideration of methods required for the prevention of plant diseases that may affect the quality and yields of essential food crops.

Everything that is made from wood, cloth to hide our nakedness, coal, oil, and gasoline to warm our homes, to generate power, and to propel transportation vehicles, paper on which to record the history of mankind, and most human and animal foods originate directly or indirectly from growing plants. In addition to these practical uses, plants of varied hues and shapes beautify our countryside, and brighten up our homes and gardens. Anything that destroys or impairs plant development, whether it be farm crops, trees of the city parks, or flowers in the home garden, may affect the individual and in some cases has caused national hardship and privation. Henry T. and Rebecca T. Northen (1954), in their interesting book, "The Secret of the Green Thumb," have given a readily understandable description of the nature of living plants:

> A man-made machine, although it is designed to work automatically, is subject to all kinds of failures. A switch may break, a line may become disconnected, a bearing may wear out, or any number of troubles may interfere with its working order. Power failure or lack of fuel will put it completely out of operation. A plant puts to shame any such creation of man. It locates itself in the midst of the raw materials it needs, builds and repairs its own machine, and makes its own fuel. The raw materials it takes in and the fuel it makes are distributed to all of its parts by a two-way system of supply lines. It can never suffer from a power shortage as long as the sun is in the sky, for it operates on solar energy, which incidently puts it one step ahead of the scientists who have harnessed atomic energy. This wonderfully efficient machine is held in the deceptively simple structure of roots, stems, and leaves. If man wants to move plants from their native habitats to have them for himself, if he wants to place them at will to suit his aesthetic tastes, he must consider what they require from their environment and how they function in it.

Plants have survived through the geologic ages and have persisted in the smog-laden environment of man-made industrial cities. They are among the oldest species of life. Despite their antiquity and unique ability to adapt to unnatural environments, most plants are subject to a wide spectrum of maladies and disorders which have become the subject of study, research, and resolution by the science of phytopathology.

Phytopathology, sometimes referred to as "plant pathology," is the modern science concerned with the identification, causal agents, and control of disorders that impair the usefulness or mar the beauty of agronomic crops, trees, shrubs, flowers, fruits, vegetables, and other members of the plant kingdom.

The word "phytopathology" is of Greek origin, being derived from "PATHOS" meaning disease, "PHYTON" meaning plant, and "LOGOS" meaning discourse. Thus, literally, phytopathology means "the science of discourse, or discussion, concerning plant diseases."

Horsfall and Dimond (1959) define plant pathology as "the art of treating the sick plant, and the science of understanding the nature of the diseased plant. Art is doing, science is understanding. Hence, Plant Pathology is an applied science which means a science that should be put to work."

The late Elvin C. Stakman, internationally respected scientist, Dean of American Phytopathologists, and major professor of many internationally recognized plant disease specialists who were trained by "Stak" at the University of Minnesota, once stated:

> For better or for worse, Plant Pathology has its genesis in fields and granaries rather than in Halls of Ivy. The emphasis on the practical is nothing to be ashamed of. On the contrary, plant pathologists can be proud they are needed by society and not merely tolerated.

WHAT IS A PLANT DISEASE

If a logical discussion of plant disease control with chemicals is to be compiled, it becomes necessary to first resolve exactly, what is a plant disease? Many definitions of plant disease are given in the standard textbooks dealing with plant pathology. The exact definition of what a plant disease includes has always been a subject for argument among many professional plant pathologists and universal agreement has still to be attained.

To the layman a plant disease is "any abnormality of a plant, its parts or products that reduces the economic value or aesthetic quality." By such a definition injury from frost, drought, lightning, mineral deficiency, or mechanical damage would have to be included in the definition of disease. To the layman, a plant is either sick or healthy.

Stakman and Harrar (1957) defined a plant disease as follows:

> The term "Plant Disease" is properly applied to any deviation from normal growth or structure of plants that is sufficiently pronounced and permanent to produce visible symptoms, or to impair quality and economic value.

Thus, according to this definition, mechanical damage, nutritional deficiency, chemical injury, and adverse effects of the environment are responsible for the initiation of disease in plants.

George N. Agrios (1969), of the University of Massachusetts, provided a more modern and technical definition of plant disease, as follows:

> The expression of plant disease is the sum of biochemical reactions taking place at the point of contact of the pathogen with the host cells, or at the point of reaction of the enzymes, toxins etc., secreted by the pathogen within the host cells. A whole plant, however, may become diseased by certain pathogens when only a small area involving the main conductive tissues of the plant are invaded and translocation failure results. Although man recognizes plant diseases by the symptoms, or the unhealthy condition of affected plants, it should be remembered that diseases are not symptoms or conditions, nor are they the instigators of the symptoms or conditions, but rather they are the interractions between host and pathogen, or between host and adverse environmental conditions that result in such symptoms or abnormal conditions. Thus the pathogen is not the disease—it is the cause of the disease. DISEASE IS THE SUM OF THE NORMAL CHEMICAL REACTIONS THAT ARE INHIBITED, OR THE ABNORMAL CHEMICAL REACTION INDUCED INSIDE THE CELLS AND IN THE TISSUES OF THE PLANT AS A RESULT OF THE IRRITATION BROUGHT ABOUT BY THE CAUSAL AGENT.

George L. McNew (1960), of the Boyce Thompson Institute, Yonkers, New York, defines a plant disease as:

> Any interference with the normal metabolism of a plant that prevents it from making full use of the factors available in its environment for growth and reproduction must be considered as a disease. The agent that interferes with the physiological processes of a plant is a pathogen regardless of whether it is living or inanimate.

The U.S. National Academy of Science in 1968 defined a plant disease as:

> "An alteration in one or more of the ordered sequential series of physiological

processes culminating in a loss of coordination of energy utilization in a plant as a result of the continuous irritations from the presence or absence of some factor or agent." In other words a harmful irritation of the normal physiological and biochemical development of the plant.

The National Academy of Science's definition would also include macro and micro deficiencies in nutrient requirements, herbicide injury, mechanical injury, and adverse environmental effects as plant diseases.

From the standpoint of plant disease control with fungicides the definition of a plant disease outlined by Clare Kenaga (1974) of Purdue University would seem to more clearly delimit the zone of operation for chemical control of plant diseases. Kenaga has defined plant diseases as either "infectious" or "non-infectious" plant diseases. A non-infectious disease is defined as "an abnormality in plants caused by agencies such as nutrient deficiencies, weather conditions, mechanical injury, chemical injury, and damage by hail, wind, frost, lightning, and other external factors which are non-infectious, and their effects are not transmissible from affected plants to healthy plants." An infectious disease is "an abnormal plant condition caused by biological agencies such as fungi, bacteria, viruses and nematodes and are capable of being transmitted from diseased plants to healthy plants to cause disease on the latter under favorable environmental conditions."

The agent responsible for initiating an infectious plant disease is referred to as "THE CAUSAL ORGANISM, THE CAUSAL AGENT OR THE INCITANT," implying that it may incite a disease condition under the influence of other factors.

In this consideration of chemicals for plant disease control the Kenaga definition of plant disease is preferred because the plant protectants discussed are used for PROTECTING THE HOST AGAINST PHYTOPATHOGENS CAPABLE OF INCITING INFECTIOUS PLANT DISEASES. The term "fungicide" literally means any agent capable of killing fungi. However, by common usage, the term "fungicide" has come to include protective or eradicative agents for the suppression or control of infectious plant diseases of fungal and bacterial etiology. Physical agents such as heat, ultraviolet light, radioactive elements, steam sterilization, and other physical agents may in some instances be referred to as fungicides, although it is preferable to refer to such methods for plant disease control as "physical methods for plant protection."

Chemicals which inhibit, inactivate, or destroy the ability of fungi to grow and reproduce are correctly referred to as "fungicides." Chemicals that temporarily inhibit the growth of fungi without destroying their capability to reproduce are more correctly referred to as "Fungistats." Chemicals that prevent the spread of a plant disease by inhibiting sporulation of the causal agent are referred to as "anti-sporulants."

To more effectively discuss the uses of chemicals for the control of infectious plant diseases, a brief discussion of the nature of plant diseases is desirable. A biological inciter of infectious plant disease is only infrequently the sole cause of abnormal plant conditions, and in most instances it must operate in conjunction with other ambiotic factors that influence the host-parasite relationship, ultimately causing plant disease symptoms.

The classical representation of the interacting factors necessary for the occurrence of an infectious plant disease is illustrated in Fig. 1.3. Einar Palm (1972), of the University of Missouri, has described parasitic diseases as follows:

> Parasitic diseases are caused by *living* agents that enter into a dynamic relationship with the host plant. They can multiply and be disseminated from plant to plant. Wind and water movements are responsible for the dissemination of many plant disease organisms. Insects or other vectors are also responsible for transmitting many disease organisms.

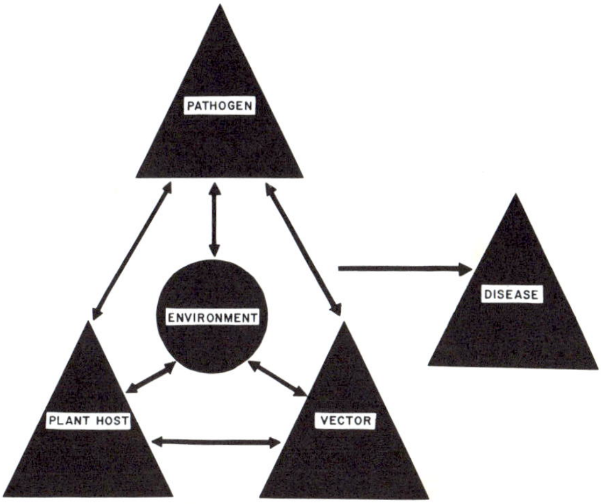

FIG. 1.3. INTERACTING FACTORS REQUIRED FOR PLANT DISEASE DEVELOPMENT

The principle groups of infectious agents (pathogens) include the (1) fungi, (2) bacteria, (3) viruses, (4) mycoplasmas, (5) nematodes, and (6) parasitic plants.

The conditions necessary for a plant disease to occur are (1) a susceptible host plant, (2) a disease-producing agent, pathogen, that can enter into a specific host-pathogen interaction, and (3) an environment that is favorable to the development of the disease. If any of these conditions are not met, a disease probably will not occur. The interaction of these three conditions is known as "the disease complex" (Fig. 1.4). The brevity of this definition should not lead to over-simplification in evaluating plant diseases. All parts of the disease complex are extremely complicated because they are all affected by innumerable biological and environmental variables.

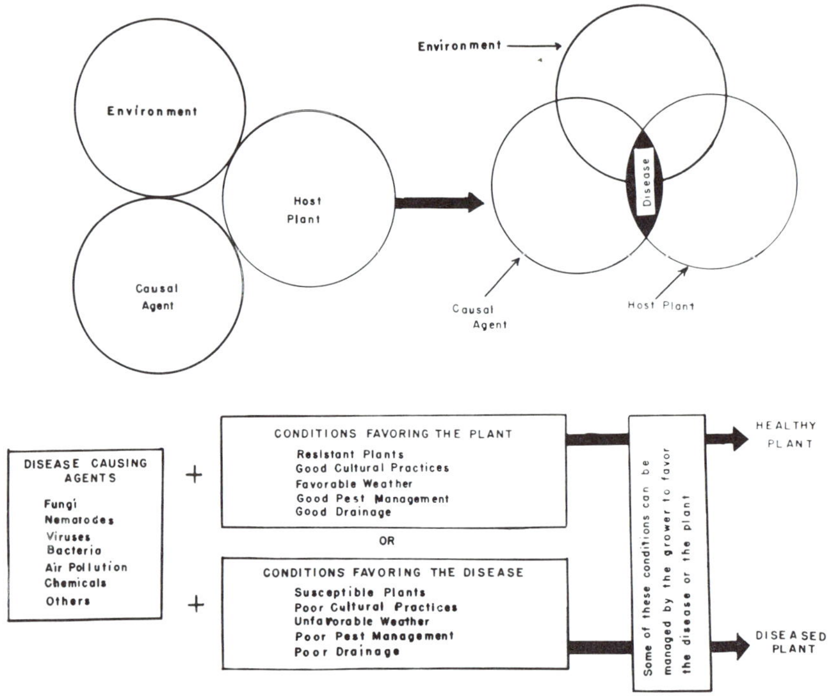

Adapted from Palm (1972)

FIG. 1.4. THE NATURE OF PLANT DISEASES

Some people have the mistaken idea that biological agencies such as fungi, bacteria, and viruses can by themselves cause phytopathological problems. This is not so, since phytopathogens, the incitants of the infectious plant disease situation, must occur in association with a susceptible host, a favorable environment for the infection processes, and a vector or agent for the transmission and spread of the causal agent. When all of these factors are present and in balance, a plant disease may then begin and disease symptoms may subsequently develop.

Galli *et al.* (1968), in their (Brasilian) Manual de Fitopatologia—Doencas das Plantas e seu Controle, have uniquely described and illustrated the cycle of host-pathogen relations of infectious plant diseases

(Fig. 1.5). According to this source, for an infectious plant disease to be initiated there has to be a source of inoculum from natural sources of the pathogen capable of inciting disease. This can be from soil-borne overwintering structures of the phytopathogen, internally or externally seedborne inoculum, air-borne spores from other areas, inoculum from local native hosts, or from local contaminated crop residues. In order for inoculum to reach the infection court it must be disseminated or carried to a susceptible host. This can be accomplished by insect carriers, by air currents, by rains, by surface water, or by introduction of infected plant material. After the inoculum of the disease-inciting pathogen has been distributed and has reached a susceptible host, then environmental conditions of temperature and humidity and host susceptibility must be favorable for the INOCULATION CYCLE, which is comprised of germination, penetration, and colonization of the pathogen within the host. When this has occurred, SYMPTOMS OF THE DISEASE, or disease expression, will be observed on the SUSCEPTIBLE HOST. At this time REPRODUCTION of the pathogen will occur, completing the PRIMARY CYCLE of the HOST-PATHOGEN RELATIONSHIP.

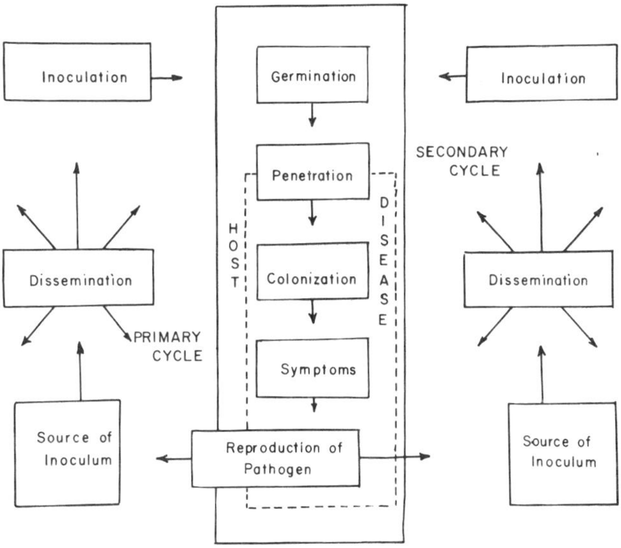

Adapted from Galli et al. (1968)

FIG. 1.5. HOST-PATHOGEN RELATIONS OF INFECTIOUS PLANT DISEASES

If environmental conditions are favorable, the SECONDARY INFECTION CYCLE will be initiated. Again, this requires a natural source of inoculum, dissemination of the inoculum, and favorable conditions for secondary infection. It is the secondary cycle of host-parasite relationships which is generally responsible for the yield-reducing results of plant diseases.

For those interested in a more complete description of the nature of plant diseases, their causal agents, and factors affecting their development, suggested references are listed on p. 14-15.

For many years the study of plant diseases was an avocation of the medical profession. With the proof by Anton de Bary that fungi were implicated as one of the causes of plant diseases, plant pathology was conceived and born. Thus, in the U.S.A. the science of plant pathology is less than 100 years old and dates from the establishment by the U.S. Department of Agriculture of the Division of Plant Physiology and Vegetable Pathology in 1885. Today, departments of plant pathology are included in most of the land-grant universities in the U.S.A. and in many of the developed countries of the eastern and western hemispheres.

The science of Applied Plant Pathology is one of the youngest of the agricultural sciences. In general, effective practices for plant disease control are imperfectly understood by some laymen, by some practical agriculturalists, and even by some university and experiment station workers.

In recent years the activities of groups such as some bird watcher clubs, organic gardening societies, wildlife conservation groups, and some federal agencies have created confusion and distrust in the use of chemicals for plant disease control and for the suppression of other unwanted pests. So extreme have some of these publicized claims become that mother's milk, hair dyes, artificial food sweeteners, nail polish, and other common commodities have been suggested as contributors to the incidence of human cancer. Distorted publicity and emotional debate have convinced a significant segment of society that chemicals for plant disease control, like some of the products used for insect control, are "poisons on the family dining table," and for this reason should be completely eliminated from the agricultural food production industry.

In this compilation of principles and practices involved in the use of chemicals for crop protection against destructive plant diseases, an attempt has been made to draw aside this curtain of confusion and misunderstanding, to separate facts from fiction, and to provide useful and unbiased information concerning the safe use of fungicides for the reduction of food losses caused by plant diseases.

BIBLIOGRAPHY

AGRIOS, G.N. 1969. Plant Pathology. Academic Press, New York.

GALLI, F. *et al.* 1968. Manual de Fitopatologia. Doencas das Plantas e seu Controle. Biblioteca Agronomica Ceres. Sao Paulo, Brazil.

HORSFALL, J.G., and A.E. DIMOND. 1959-1960. Plant Pathology. An Advanced Treatise. 3 volumes. Academic Press, New York.

KENAGA, C.B. 1974. Principles of Phytopathology. Second edition. Balt Publishers, Lafayette, Indiana.

MCNEW, G.L. 1960. The nature, origin and evolution of parasitism. *In* Plant Pathology, Volume 2. J.G. Horsfall and A. Dimond (editors). Academic Press, New York.

NORTHEN, H.T., and R.T. NORTHEN. 1954. The Secret of the Green Thumb. The Ronald Press, New York.

PALM, E.W. 1972. What you should know about plant diseases. Univ. of Missouri Ext. Mimeo. Columbia.

STAKMAN, E.C., and J.G. HARRAR. 1957. Plant Pathology. The Ronald Press, New York.

WALKER, J.C. 1957. Plant Pathology. Third edition. McGraw-Hill Book Co., New York.

2

General Principles of Plant Disease Control

The reduction or prevention of crop losses caused by infectious plant diseases assumes increasing importance as world populations continue to increase. The task of providing sufficient food to permit adequate diets for steadily rising human populations is becoming increasingly difficult. This is especially true in some of the developing nations.

According to USAID (U.S. Agency for International Development), the world population is growing at the rate of 2.6% yearly, and by 1985 it is estimated that the total world population increase will amount to 2 billion extra people. It has been predicted that two-thirds of this increase will occur in less developed nations in the tropics and subtropics. The Food and Agriculture Organization has suggested that by 1985 world food production levels must increase by 140% over 1965 production levels to avoid the specter of recurring widespread famines. Obviously, in each tropical and subtropical nation, food production must equal population growth to prevent a worsening of the present food situation. A still larger increase will be needed to alleviate present areas of distress in the available amounts of food, and to produce diets that are much better balanced in proteins, minerals, and vitamins. Further substantial increases in crop production will be needed to provide surpluses that will permit the small farmer to enter the cash economy, to make possible his acquisition of better clothing, housing, sanitation, medical services, education, and communications. The need for increased world food production is obvious, and the reduction of losses presently caused by plant diseases is one approach to the amelioration of the world food problem.

On the home front, plant disease losses cause little concern in times of plenty. However, with increasing production costs, decreased farm profits, and increased cost of living standards, crop losses from plant diseases such as southern corn leaf blight, apple scab, and wheat rust make the difference between profit and loss for the professional farmer, and increase costs for the consumer.

For the home gardener who learns with experience that there are hazards to growing plants for pleasure, plant disease and insect controls are important items in the home gardening plan, and their omission will inevitably result in disappointment or even failure.

The control of parasitic plant diseases is the ultimate objective of the science of plant pathology, and is predicated by an understanding of the nature of plant diseases and the influence of the environment on their development and severity.

Plant pathology, the youngest of the agricultural sciences, has developed from the lowly status of a minor segment of botany to a position of significance as a partner with other sciences in the world food production industry. However, plant pathology cannot be divorced from related sciences, and the methods for plant disease control are based in large measure on principles and practices adopted from related sciences, applied to the specific task of reducing or preventing losses caused by parasitic plant diseases.

Before outlining the general principles of plant disease control it is important to describe and understand the relationship of plant pathology to other sciences in agricultural technology. This relationship is illustrated in Fig. 2.1. In this wheel of crop production and plant husbandry there are ten basic sciences that are the principal components of a successful crop production program. Thus, the wheel of crop production and successful home gardening may be depicted as a multispoked wheel. When all of the spokes are fitted into the hub of the wheel, maximum efficiency will result. When any one of the essential spokes is omitted, the wheel could collapse with resultant loss and probable disaster. Many examples of this type of situation could be cited. For example, the southern corn leaf blight epidemic of 1970-1971 was a major disaster in the principal corn producing areas. In this case the spoke of genetics and plant breeding was faulty. Corn hybrids containing Texas male-sterile cytoplasm became highly susceptible to the southern corn leaf blight pathogen (*Helminthosporium maydis*). In this case the disease resistance spoke collapsed; the result was a major economic disaster.

For the home garden rose enthusiast who has lavished TLC (Tender Loving Care) on a new, expensive, and widely advertised rose variety, dreams of an abundance of beautiful blooms will quickly evaporate when controls for aphids, black spot, and powdery mildew are omitted from the annual gardening plan. Omitting the spokes of plant pathology and entomology are frequently the most common cause for frustration and disappointment in the backyard flower garden and vegetable patch.

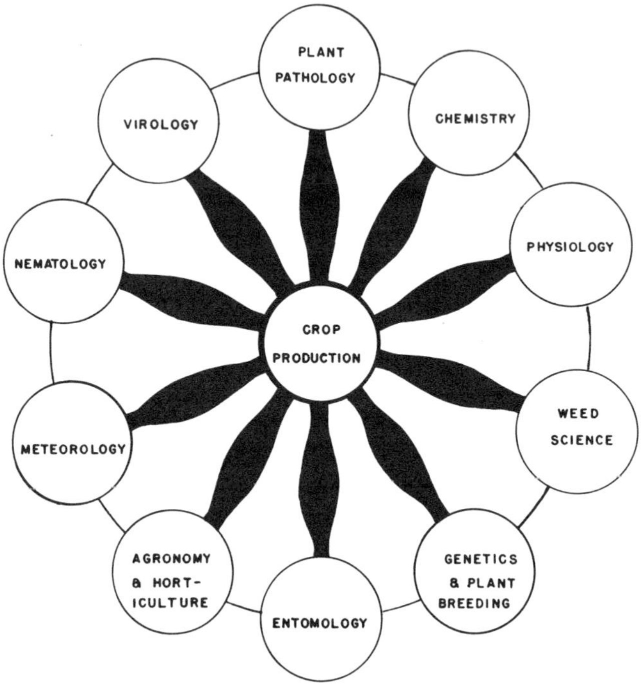

FIG. 2.1. THE RELATIONSHIP OF PLANT PATHOLOGY TO OTHER AGRICULTURAL SCIENCES

The principal spokes in the wheel of crop production, the botanical sciences (horticulture and agronomy), plant physiology, chemistry, genetics and plant breeding, nematology, meteorology, virology, entomology, phytopathology, and the relatively new science of weed science, are becoming more and more essential as crop production technology continues to advance. An appreciation and understanding of the relationship of the various sciences in crop production and home gardening are fundamental to a logical understanding of the general principles of plant disease control.

The general methods of plant disease control may be discussed by two approaches:

a. Manipulation of the plant disease triangle
b. The Immunization-Prophylaxis system

MANIPULATION OF THE PLANT DISEASE TRIANGLE

As has been discussed previously, a plant disease can occur only when all of the elements of the plant disease triangle are favorable for initiation of the plant disease complex. Thus, a logical approach to the problem of plant disease control would be to manipulate the plant disease triangle of host, pathogen, and environment in such a way as to disrupt the sequence of conditions necessary for disease establishment. This is one way of cataloging the basic methods of plant disease control (Fig. 2.2). Let us consider what may be done to the environment (or climate), the host, and the pathogen to prevent or reduce losses caused by plant diseases. First of all, what can be done to manipulate the environment to reduce plant disease severity?

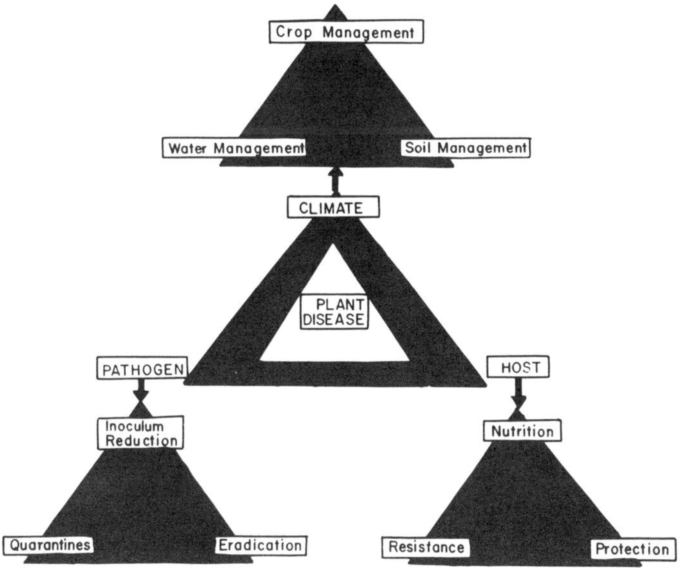

FIG. 2.2. THE GENERAL PRINCIPLES OF PLANT DISEASE CONTROL

Water Management

Under certain conditions the Crazy Top disease of corn, caused by the downy mildew fungus *Sclerophthora macrospora* Sacc., may cause serious yield reduction in field corn. The conditions necessary for this to occur are poor field drainage and waterlogged soils. Adequate tile drainage will virtually eliminate the problem. A common cause of damping-off of seedling plants, especially in home gardens, is excessive watering and waterlogged soils. Easing up on the watering schedule and light dusting of the soil surface with zinc oxide (a non-fungicide), or with a fungicide such as Captan, will help to control the problem. A condition in geraniums known as "edema," is a non-parasitic disease associated with excessive

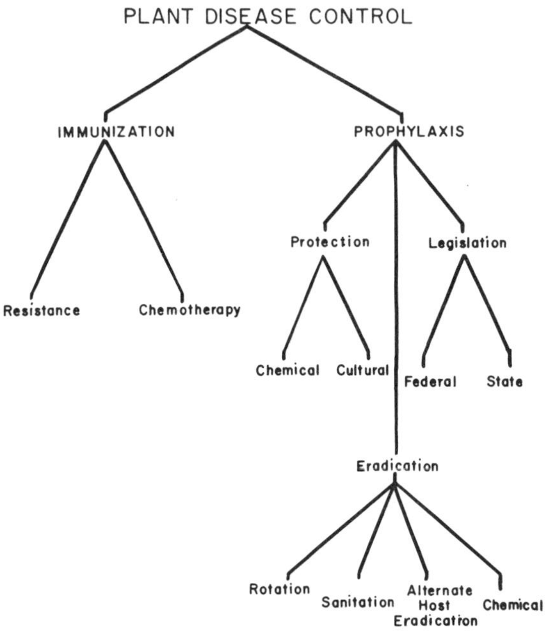

FIG. 2.3. THE IMMUNIZATION-PROPHYLAXIS SYSTEM OF CLASSIFYING METHODS FOR PLANT DISEASE CONTROL

soil moisture and retarded plant transpiration. Water-soaked, corky brown spots will develop on the leaves of affected plants, making them unsightly and resulting in dropping of leaves. If the affected plants are given good ventilation and light without excessive watering, edema will not be a problem. Excessive overhead sprinkling of garden plants, especially in the evening, is an excellent way to distribute fungus spores, resulting in rapid spread of plant disease problems. Plants such as roses, tomatoes, chrysanthemums, bluegrass lawns, green beans, lilac, peonies, and other flowering plants should be watered before dews start to develop, and soil surface watering is preferable to overhead watering of the leaves and aerial parts of plants.

Soil Management

There are several examples of the value of soil management as a method for plant disease control. Adjustment of soil pH, or hydrogen-ion concentration, has been widely used for the commercial control of the common potato scab disease (*Actinomyces scabies*). Highly acid soils are unfavorable to the causal pathogen, and growing potatoes in soils with a pH of 5.2 or below will significantly reduce losses caused by the potato scab disease. On the other hand, the club root disease of cabbage, caused by *Plasmodiophora brassicae*, is favored by acid soil reactions and causes little damage at a soil pH of 7.0 or above. Liming of the soil to elevate the soil pH has been practiced commercially for many years for the control of the clubroot disease of crucifers. Cotton wilt disease (*Fusarium oxysporium* f. *vasinfectum*) is usually most severe in potash-deficient soils and may be reduced in severity by the application of high potash fertilizers. For the home gardener, soil management practices can be used to promote more successful gardening. The use of gypsum and other commercial soil additives is practical in the home garden to improve the physical qualities of heavy soils which favor the severity of a variety of root rot and stalk rot problems. Soil management practices are an important consideration in any plant disease control program on the farm or in the backyard garden.

Crop Management

The rotation of crops is an old practice followed for many years in commercial agriculture. In the home garden, rotation of vegetables and some fruit crops is advisable to avoid severe losses from soil-borne phytopathogens. For example, tomatoes should not follow tomatoes, and string beans should not be grown in the same location year after year. Strawberry plantings should be moved to a new location every third year if possible to avoid build-up of root rot problems. Similarly, raspberries should not be replanted in the same location to avoid losses caused by raspberry anthracnose and soil-borne diseases. With the increased acceptance of minimum tillage practices in field crops, diseases such as corn stalk rots appear to increase in severity after approximately three years of continuous corn. In the same way, continued cropping of soybeans in the same location will result in increased damage from leaf blight and root rot diseases. Most modern farmers follow a crop rotation system to avoid a build-up of plant disease problems.

Sanitation practices are often neglected on the farm and in the home garden. This is frequently a direct cause of increased plant disease problems due to carry-over of pathogen inoculum on diseased plant debris. Even though minimum tillage is to be practiced with some field crops, a light disking to bury infected plant debris will greatly reduce the severity of certain plant diseases of corn, soybeans, and sorghum. Removal of all plant refuse from the home garden at the end of the growing season is just as important as sanitation in the home.

THE IMMUNIZATION-PROPHYLAXIS SYSTEM

Immunization simply means to render the host immune from attack by potential pathogens. In practical usage immunization has been employed as meaning the improvement of strains of the host through hybridization and/or selection of disease-resistant plant types. This usually is referred to as "the development of disease-resistant varieties." Immunization by genetic manipulation is usually hereditable but may be only temporary. Non-hereditable resistance may sometimes be achieved by manipulation of pre-disposing and post-infection environmental factors such as temperature, soil moisture, and fertility. Artificial immunity or resistance may also be induced by chemical means (chemotherapy), although this is still a relatively undeveloped possibility.

Prophylaxis refers to "the prevention of plant disease development." This may be achieved by "protecting" the host against infection by chemical means or by manipulation of the environment governing the disease infection process. Legislation involving the enactment of quarantines, and federal and state plant inspection services are also prophylactic control measures. Eradication of an established plant disease by crop rotation, sanitation measures, alternate host elimination, or chemical eradication are also included in prophylactic plant disease control practices.

PRACTICES DIRECTED TOWARDS THE PATHOGEN

Any practice that has as its objective suppression of plant pathogens will contribute to simplification of the task of reducing the severity of the diseases they may cause.

There are three general practices that may be directed towards phytopathogens to mitigate the disease losses they can initiate:

a. Exclusion of phytopathogens
b. Eradication of phytopathogens
c. Reduction of pathogen inoculum

Exclusion of Phytopathogens

If it were possible to completely exclude all phytopathogens from localized areas, or from countries where they do not occur, plant diseases might be of little significance. However, man has been one of the most effective means of spreading plant disease pathogens from place to place, and from country to country.

Unfortunately, several plant diseases that did not originally occur in the U.S.A. were introduced and became destructive problems before their potential for crop plant destruction was recognized.

Dutch elm disease, caused by *Ceratocystis ulmi*, has virtually eliminated most of the native American elm trees in the eastern and north-central states of the U.S.A. The disease was first recognized and identified around 1920 in Holland and several other European countries, where it caused catastrophic damage to European elms. It was not until 1930 that it was reported for the first time in Ohio on American elms, and within a period of three years it had spread to New York, New Jersey, and Connecticut. By 1940 Dutch elm disease was present in the midwest and by 1977 was causing concern as far west as California.

Investigation has established the fact that the disease had been introduced to the American continent in a shipment of veneer logs from Holland through Atlanta and Gulf of Mexico seaports. Unfortunately, some of these imported Dutch elm disease-infected logs were reshipped on railroad flatcars to veneer mills in various parts of the U.S.A.

Another example of a destructive plant disease that was introduced into the U.S.A. is the chestnut blight caused by *Endothia parasitica.* At one time chestnuts grew in the U.S.A. and were valued for their beauty and for the fruits they provided for traditional Thanksgiving Turkey festivities. In 1904 the chestnut blight disease first appeared in New York State. From this beginning spores of the causal fungus were spread by birds and air currents and the disease soon became epidemic through the Appalachians into North Carolina. Within a very few years the entire American forests of *Castanea dentata* (chestnut) were destroyed and have never again been successfully reestablished.

White pine blister rust, caused by *Cronartium ribicola*, was introduced to the American Continent on nursery stock imported from Europe. Being a heteroecious rust requiring the five-needled pine for its aecial stage, and native and cultivated Ribes species to complete the uredial and telial stages, the pathogen quickly spread to incite a destructive disease of American pines. At one time the white pine grew in extensive forests in the eastern states, the midwest, and the far west states. Susceptible species of Ribes (Currants) were also native in white pine forests and, in addition, susceptible varieties of currants were introduced for home production in the U.S.A. By 1910 the disease was reported in the New England states, in Minnesota, and in North Carolina. In the same year infected white pine nursery stock was introduced into British Columbia, Canada, and white pine blister rust disease soon appeared in destructive severity in Washington, Oregon, and northern California. Despite expenditures of federal funds for the impossible task of eradicating native and domesticated Ribes plants, the white pine blister rust disease has been responsible for millions of dollars of loss to timber stands of white pine throughout the U.S.A.

Other diseases that are known to have been introduced into the U.S.A. from other countries include the potato wart disease (*Synchitrium endobioticum*), bacterial canker of citrus (*Xanthomonas citri*), and grape downy mildew (*Plasmopora viticola*). Exclusion of the infected plant material of these problems could have prevented outbreaks of the destructive diseases referred to.

Federal Plant Quarantines

As a result of the destruction caused by introduced plant diseases, the U.S. Secretary of Agriculture proposed measures to exclude importation of infected plant material. In 1912 the United States Plant Quarantine Act was authorized by the U.S. Secretary of Agriculture with the approval of the U.S. Congress. In 1926 the Quarantine Act was further strengthened by additional legislation requiring federal certification for export of domestic plant materials. In 1957 the Federal Plant Pest Act was passed by Congress, further restricting movement of foreign and domestic plant materials.

The U.S. Plant Quarantine Act, restricting movement of plant materials within the U.S.A. and for export to other countries, is enforced by the Plant Quarantine Division of the U.S. Agricultural Research Service. Details of the Act are outlined in the 1966 U.S. Yearbook of Agriculture, "Protecting Our Food," and specific details may be obtained from the U.S. Department of Agriculture, Agricultural Research Service, Beltsville, Maryland.

Rainwater and Smith (1966) have stated, "Plant Inspection Services are maintained at 75 major ocean, Great Lakes, air and border posts of entry in the continental United States, Alaska, Hawaii, Puerto Rico, American Virgin Islands, Nassau and Bermuda." A staff of more than 500 inspectors examines cargoes of plants and plant products, and if necessary will post the required treatments or other safeguards to free them from pests. Incoming cargoes on aircraft, ships, trucks, trains, and automobiles are examined, as well as baggage of travellers and postal shipments, for compliance with U.S. Plant Quarantine Regulations. Imported plant propagative material is also treated by trained personnel at specially equipped inspection stations. In addition, a number of plant species may be held under a post-entry quarantine for observation of symptoms of plant diseases. Many other countries, including France, Australia, Austria, Mexico, and the United Kingdom, have also established quarantine regulations governing entry and export of plant materials and plant products.

Domestic Quarantines

Domestic Quarantines in several U.S. states antedate the establishment of Federal Quarantine regulations. The first domestic quarantine in the U.S.A. was enacted in 1881 to prevent the spread of the Phylloxera insect on grapes. As early as 1726–1766, Connecticut, Massachussetts, and Rhode Island had regulations requiring the eradication of barberry bushes since they were suspected of being related to destructive outbreaks of the wheat stem rust disease (*Puccinia graminis tritici*).

At the present time most states have regulations which govern the sale and distribution of plant propagative material, enforced by state nursery inspection services, seed certification agencies, and similar organizations. Most of the domestic quarantines are supplementary to federal quarantines and are enforced in cooperation with the U.S. Department of Agriculture Quarantine Regulations. In some states voluntary inspections of plant propagative material have been established to protect private interests in nurseries and seed production. The most recent domestic quarantine is the regulation governing the movement of soybeans to prevent spread of the golden cyst nematode from and into major soybean-producing areas of the U.S.A.

Much controversy has existed, and still exists, over the efficacy of plant quarantine regulations for the exclusion of destructive plant diseases. Despite the arguments pro and con regarding exclusionary measures against plant disease pathogens, they should be considered as one of the logical steps for the reduction of annual losses caused by plant diseases.

The main purpose of the U.S. Federal Plant Quarantine regulations has been to restrict the export and importation of plant pathogens from and into the U.S.A. For details on federal plant quarantines the reader is referred to Walker (1969), Dean (1953), Stakman and Harrar (1957), and Horsfall and Dimond (1959–1960).

Eradication of Phytopathogens

Eradication of phytopathogens once they are well established has been a seemingly impossible undertaking. Nevertheless, anything that can be done to reduce the inoculum of destructive phytopathogens will lighten the burden for other methods of integrated plant disease control. Large scale eradication programs for the elimination of established plant diseases have succeeded as major control measures only infrequently. The Dutch elm disease eradication program during the 1930's may have slowed down the spread of the disease but it failed in areas where the problem was already established.

A classical example of a successful plant disease eradication program was the citrus canker eradication program in Florida. The citrus canker disease, caused by the bacterial phytopathogen *Xanthomonas citri*, was first observed and identified in Florida in 1913. Two years later an eradication program was started in several citrus-producing states. Eventually, seven states and the U.S. Department of Agriculture cooperated in a program to eradicate citrus canker disease. Within two years of the program's initiation, Florida was relatively free from citrus canker disease. It was estimated that 3 million infected trees were eradicated in Florida and adjacent Gulf Coast states. Although isolated outbreaks of citrus canker have occurred in Florida since that time, no reports of major outbreaks of the disease have been reported since 1928 in the U.S.A. Mass eradication followed by continuing surveys and elimination of infected citrus trees is a classical example of a successful plant disease eradication program.

Flag smut disease of wheat, caused by *Urocystis tritici*, is generally considered to have been introduced into the U.S.A. from Australia. It became established in wheat-growing areas in Illinois, Kansas, Missouri, Oregon, and Washington. Introduction of resistant wheat varieties, supplemented by local burning of infected areas, evidently eradicated the disease since it has not been reported as a serious wheat disease in the Northwest in recent years.

Eradication of alternate hosts of rust disease pathogens has been practiced for many years. Extensive eradication of the common barberry, *Berberis vulgaris*, the alternate host of wheat stem rust (*Puccinia graminis tritici*), was organized in 18 states of the northern wheat-producing areas of the U.S.A. in 1918. According to Stakman and Harrar (1957), the federally sponsored barberry eradication program resulted in the destruction of one-half billion barberry bushes and was estimated to have reduced losses from wheat stem rust disease in the U.S.A. from an annual average of 40 million bushels of wheat to less than 15 million bushels at a total cost of less than 1% of the value of the wheat, oats, barley, and rye crops in the 18 states involved in the eradication program. Stakman and Harrar (1957) have estimated that the barberry eradication program resulted in annual savings for U.S. wheat farmers in excess of $30 million.

Another well known alternate host eradication program was the attempt to eliminate wild species of *Ribes* which are the alternate hosts of white pine blister rust (*Cronartium ribicola*). It has been estimated that during the 1920's more than one and one half billion *Ribes* bushes were eradicated from the vicinity of white pine forests, with a subsequent reduction in timber losses caused by the pine blister rust disease.

In many areas the replacement of red cedar windbreaks with other ornamental shadetrees has reduced commercial apple orchard losses from the cedar rust disease (*Gymnosporangium juniperi-virginianae*) for which the red cedar is an alternate host.

Stakman and Harrar (1957), in the conclusion of their section on eradication programs for plant disease control, have stated:

> An examination of the record suggests that large-scale eradication falls far short of being an ideal method of controlling plant diseases. In the control of citrus canker it has been strikingly effective, but in many cases results have been inconclusive at best, and in others eradication efforts have been unsuccessful. Among the most successful eradication campaigns have been those designed to eliminate the alternate hosts of heteroecious rusts.
>
> Eradication programs applied to economic hosts are always expensive, in terms of both the cost of the campaign and the losses in crop production. Those which have occurred in the past are the best evidence of the dangers inherent in introduction of exotic pathogens and the necessity for control programs. It is to be hoped that such introductions can be avoided in the future or that, if they do occur, prompt action based on scientific information will permit complete eradication before major pathogens become permanently established.
>
> In a sense, ex-post-facto quarantines and eradication programs are negative measures for the control of epidemic plant diseases; action is taken after initial damage is done. Positive control measures are those which prevent combinations of circumstances that permit development of epidemics or which protect host plants from damage by pathogenic microorganisms. Unfortunately, the ideal situation in which positive control can be expected to eliminate epidemics is far from being realized. It is therefore essential to use the technic of quarantine, whenever this measure offers practical possibilities, to limit the distribution of virulent pathogens. When quarantine measures fail, eradication campaigns may be the only solution to outbreaks of epidemic diseases.

Inoculum Reduction

One of the most frequently overlooked methods for reducing the severity of plant diseases is the use of various practices that may be followed to reduce inoculum (infection sources) of phytopathogens capable of initiating plant disease infections. It is often surprising to see an immaculately maintained home with an untidy backyard garden wilderness overgrown with weeds and dead plant remains. Cleaning up the backyard garden in the fall and spring has as its objective not only the removal of unsightly plant debris from the past growing season, but also removing "carry-over inoculum" of plant disease pathogens that can initiate plant diseases again during the coming season. This is commonly referred to as "sanitation." A few hours in the fall or early spring devoted to raking up black spot-infected rose leaves, scab-infected apple leaves, removal of anthracnose-infected raspberry canes, and the burning of old disease-infected grape, melon, cucumber, tomato, and potato debris will significantly reduce sources of plant disease infection for the coming season.

Thus, the removal of "hold-over cankers" is effective in reducing the severity of the fireblight disease of apples and pears (*Erwinia amylovora*). A common practice of commercial strawberry growers is to mow the leaves from berry plantings after harvest, raking them up and burning them to reduce the carry-over of strawberry leafspot inoculum (*Mycosphaerella fragariae*). This will significantly reduce the severity of common strawberry leafspot the following season.

In the home flower garden the collecting and destruction of blossoms infected with graymold (Botrytis) is a practical method for reducing damage caused by this problem in roses, peonies, tulips, and other flowers in the home garden. The removal of brown rot-infected plums, cherries, and peaches from commercial packing sheds during the harvest season and disposal in remote refuse dumps will reduce post-harvest, in-transit losses from brown rot (*Monilinia fructicola*). Since spring infections of brown rot blossom blight of peaches may originate from conidia (asexual spores) of the brown rot fungus produced from overwintered mummies on the trees, removal of these by the amateur grower will contribute towards control of the problem later in the season. In a similar way, removal of the jelly-like "cedar apples" from red cedar trees in the spring will reduce the severity of cedar apple rust disease on susceptible apple varieties.

General sanitation practices such as removal of infected plant debris, removal of dead and diseased branches from bearing fruit trees and valued shade trees prior to the growing season, sterilizing of knives used for cutting potato seed pieces, roguing out virus-infected raspberry plants, and similar practices are of value in reducing the severity of several common plant diseases. Many other examples of reduction of plant disease losses by common sense sanitation practices could be cited. While reduction of plant pathogen inoculum usually will not completely control a specific disease problem, it will in most cases lighten the burden of the Battle of the Blights during the growing season.

PRACTICES DIRECTED TOWARDS THE HOST—PLANT NUTRITION

Since the presence of a susceptible host—the victim—is necessary for the establishment of a plant disease, it is obvious that numerous factors will govern or condition host response to potential infection by a pathogen. Here again, the environment plays a vital role in the host-disease complex, and offers another opportunity to manipulate the plant disease triangle to prevent or restrict plant disease development. As is the case in human and animal disorders, vigor of the host will generally influence disease susceptibility. Usually, but not always, a vigorous plant is a healthy plant, and is frequently more resistant to attack by phytopathogens than a nutritionally deficient plant.

Plants, like animals, need a balanced diet to make normal growth. When nitrogen is deficient, growth is generally slowed down or stunted. In humans and animals, iron is essential for adequate formation of hemoglobin and its deficiency may result in anemia. Plants are similar to animals in this respect and when iron is deficient plants also may become anemic since iron is essential for chlorophyll formation. Plants deficient in iron may develop yellow or chlorotic leaves, which is similar to anemia in animals and humans. Deficiency in calcium or phosphorus in animals may result in poor bone structure. Potash deficiency in plants may also result in poor structural development and increased susceptibility to stalk and root rot diseases.

There are 15 essential elements necessary for normal plant development and absence of some of these may influence susceptibility to destructive plant diseases. Three of the fifteen essential elements, oxygen, carbon, and hydrogen, are derived from air and water. The remainder, nitrogen, phosphorus, potassium, calcium, magnesium, iron, and sulfur, are acquired as dissolved mineral salts from the soil, and are referred to as "macro elements." The trace elements manganese, boron, copper, zinc, and molybdenum are also acquired as dissolved salts from the soil and are referred to as "micro elements."

In general, the balance of plant nutrients may be more important to plant health than concentrations of total fertilizer nutrients when plants are exposed to attack by plant pathogens. A deficiency or surplus of any of the macronutrients may often favor disease development. No general rules can be given for plant disease reduction by fertility practices. Each disease must be considered individually. A few specific examples of the influence of plant nutrition on disease control are briefly described.

Blossom-end rot of tomatoes will frequently cause considerable fruit damage and loss early in the season. This can usually be avoided by a uniform supply of moisture and applications of additional calcium. It is a common practice to side-dress tomatoes with readily available nitrogen fertilizers preceded with an application of phosphorus and calcium to prevent losses from blossom-end rot.

In small grains such as wheat, oats, and barley, heavy applications of nitrate promote rapid growth and increase susceptibility to powdery mildew and leaf rusts. Addition of potash to balance excess nitrogen will tend to lessen the severity of these diseases. Excess nitrogen has long been recognized as predisposing apples and pears to increased damage from the fireblight disease.

In cotton, susceptibility to bacterial blight (*Xanthomonas malvacearum*) is increased by nitrogen deficiency, and slightly susceptible varieties will become resistant when provided with an adequate supply of nitrogen. Susceptibility to Verticillium wilt, on the other hand, is increased by an excess of available nitrogen. The same situation also applies to the severity of the cotton boll disease. Here again, balanced fertility is necessary in the overall cotton disease control program.

Plants which are deficient in certain micro-elements may become more susceptible to certain diseases. Sulfur deficiency in onions, calcium deficiency in tobacco and tomato, boron deficiency in alfalfa, and magnesium deficiency in citrus and soybeans may result in the development of undesirable physiological disorders.

The recent research of Huber and his associates on nitrification inhibitors suggests a unique possibility for further utilization of plant fertility in the control of some plant diseases. Nitrogen's influence on nutrition as related to plant disease severity has been a subject for extensive research (Huber, Warren, Nelson, and Tsai 1977; Huber and Watson 1974). Nitrogen is the element required in the greatest amount for productive plant growth; nitrogen has limited availability in most soils but has a major influence on cell size, cell wall development, and the plant's chemical constituents. Huber has emphasized that the form of nitrogen taken up by plants can influence the severity of some plant diseases, thus providing an opportunity to reduce disease severity by using nitrification inhibitors to manipulate crop production without sacrificing the nutritional status of the crop. The work of Huber demonstrated that the effect of a specific form of nitrogen on disease severity when taken up by plants depends on rate and timing of soil nitrogen applications, previous cropping systems, and the prevailing physical environment. The interacting effects of these factors were illustrated by the reduction of losses from root and foot rots of cereals caused by the fungus *Ophiobolus graminis* when inorganic ammonium nitrogen ($NH^{+}{}_{4}$-N) is absorbed by cereal plants, but is increased by take up of nitrate nitrogen (NO_3-N). The work of Huber suggests a major possibility for using non-pesticidal chemicals in integrated plant disease control programs.

Consideration of the nutritional requirements for growing plants has been largely overlooked as a contributing factor for integrated plant disease control.

PLANT DISEASE RESISTANCE

It has been known for some time that plant species differ in their susceptibility to plant diseases, and probably some plant types have survived because of these attributes. In the early part of the 19th Century it was observed in England that certain wheat varieties were resistant to leaf rust. The Rev. Miles Joseph Berkeley in 1851 observed and reported that white varieties of onions were attacked by the onion smudge disease (*Colletotrichum circinans*) while colored varieties growing alongside were free from the disease.

The application of genetic breeding for resistance in the control of plant diseases has been the most significant and profitable approach to the problem of reducing losses caused by destructive plant diseases. There are some authorities who believe this is the "ultimate" in plant disease control and the only approach worthy of serious scientific consideration. Nevertheless, practical considerations indicate that at the present time the millenium of total plant disease control by means of host resistance has not yet dawned. It must be recognized that the development of disease resistant crop varieties, and their introduction into commercial agriculture has been an important cornerstone in the advancement of modern agricultural crop production. However, it is not the only practical approach to applied methods of plant disease control.

For the layman there is sometimes confusion as to the meaning of plant disease resistance. Obviously, there are several types of disease resistance which are not synonymous with immunity.

According to Kenaga (1974), disease tolerance in plants may be defined as, "the ability of the host plant to survive and produce satisfactory yields at a level of infection that causes economic loss to other varieties of the same species." The plant is then said to be DISEASE TOLERANT. Examples of plant disease tolerance are wheat varieties that may be severely infected with leaf rust (*Puccinia recondita tritici*) and still produce economically satisfactory wheat yields.

APPARENT RESISTANCE, or disease escape, may be defined as the capacity of a susceptible plant to avoid infectious diseases through some character of the plant, or other factor, that prevents infection. Plants possessing the capacity of disease escape are still susceptible. Therefore, if a change of conditions renders the escape mechanisms inoperative, the plant can become diseased.

STATIC RESISTANCE refers to mechanisms that are continuous on or in the host, regardless of whether the pathogen is present or not. Such mechanisms restrict or inhibit the entrance of pathogens when they attempt to enter the host directly, through natural openings, mechanical wounds, or insect injuries. Examples of this type of resistance are thickness of cuticle, natural content of chemical compounds within the host that inhibit pathogen development, and the ability of the host to heal over natural openings or mechanical injuries.

DYNAMIC RESISTANCE refers to mechanisms that are initiated by the host in the presence of a pathogen. Examples of this are hypersensitivity, development of cork tissue "walling off" the pathogen, and the development of so-called "collars" by wheat as a result of infection by powdery mildew (*Erysiphe graminis* var: *tritici*).

IMMUNITY may be defined as freedom from attack and injury by an infectious pathogen. Immunity means COMPLETE DISEASE RESISTANCE.

The concepts and application of plant disease resistance have recently been summarized in R.R. Nelson's textbook, "Breeding Plants for Disease Resistance."

The science of breeding for plant disease resistance was stimulated by the rediscovery of Mendel's Law in 1900. R.H. Biffen, in 1912, was the first to apply Mendel's law to the inheritance of plant disease resistance. Biffen crossed the wheat variety 'Rivet' with the variety 'Michigan Bronze' to produce a hybrid highly resistant to stripe rust disease (*Puccinia striiformis*). Since that time breeding for plant disease resistance has become the most widely used approach to the control of plant diseases. Nelson *et al.* (1973) estimated that 75% of current agricultural acreage in the U.S.A. is planted with varieties of crops resistant to one or more diseases.

Coons estimated in 1953 that 50% of all crop acreage in the U.S.A. utilized disease-resistant crop varieties. Compounded the world over today, the benefits accruing from the use of disease-resistant crop varieties certainly would amount to many billions of dollars annually.

The late distinguished plant breeder and phytopathologist, Ralph M. Caldwell, pointed out that nature has long been involved in breeding for plant disease resistance. One of nature's foremost breeding accomplishments has been the development of corn resistant to leaf rust caused by *Puccinia sorghi.* Caldwell (1966) described this as "one really smashing success, since there have been no reports of epidemics of corn leaf rust, even though open pollinated and hybrid corn varieties are known to be partially susceptible as proven by artificial inoculation."

Numerous examples of successful breeding of plants for plant disease resistance are cited by Caldwell (1966). Probably one of the most outstanding examples has been the breeding of small grains resistant to stem rust (*Puccinia graminis*). Craigie (1964) calculated the gain in yield and income from the use of rust resistant varieties of wheat in Manitoba and Saskatchewan, Canada, during 1938-1943, and reported the average annual gain to Canadian wheat growers to be $27 million, with an all-time total expenditure for rust research and wheat breeding by the Canadian government to be less than $2 million. U.S. losses from crown rust of oats (*Puccinia coronata avenae*) caused serious reduction in the commercial production of oats. The introduction of 'Clintland', the rust resistant variety by Caldwell and his co-workers at Purdue University, significantly reduced losses from this disease. Caldwell reported gains in Indiana of 9.5 million bushels worth approximately $6 million, which exceeded by several times the estimated investment for all research and breeding work in Indiana by State and Federal Government funds.

Efforts to produce crop varieties resistant to nematodes have also been successful, and Kehr (1966) lists 22 crop varieties resistant to one or more parasitic nematodes. Immunization by the use of disease resistant plant varieties is a major tool in the control of many plant diseases. However, it must be considered as only one of the methods of preventing losses from some plant diseases to be used in an integrated program with other methods.

DIRECT HOST PROTECTION

Direct protection of the host against invasion by phytopathogens usually refers to the use of protective chemicals known as "fungicides." Since this is the main topic of this textbook, it will be discussed in detail in the following chapters.

However, there are other ways in which the host plant may be protected against infection by disease inciting pathogens. Insecticides are generally applied for the control of insect pests. They may also be used in several instances to protect plants against insect carriers of plant pathogenic fungi, bacteria, and numerous destructive virus diseases. The use of herbicides indirectly may protect economic crops by eliminating unwanted plants that harbor phytopathogens. In addition, herbicides are used commercially to eliminate the foliage of diseased plants prior to harvest, thus reducing sporulation of fungi that may be responsible for postharvest disease problems. For example, potato vines can be killed with herbicide prior to harvest of tubers with the secondary objective of preventing infection of tubers by spores of the late blight fungus.

Direct protection of a susceptible plant host by the application of fungicides is one of the important phases of any plant disease control program which must be integrated with other control practices discussed previously.

BIBLIOGRAPHY

BIFFEN, R.H. 1912. Studies in inheritance of disease resistance. J. Agric. Res. 4:421-429.

CALDWELL, R.M. 1966. Advances and challenges in the control of plant diseases by breeding. Pest control by chemical, biological, genetic, and physical means. U.S. Dep. Agric. ARS Publ. *33-110.* Washington, D.C.

COONS, G.H. 1953. Breeding for Resistance to Disease. U.S. Dep. Agric. Yearb. Agric. Plant Diseases 174-192. U.S. Govt. Printing Office. Washington, D.C.

CRAIGIE, J.H. 1964. Increase in production and value of the wheat crop in Manitoba and Eastern Saskatchewan as a result of the introduction of rust resistant varieties. Sci. Agric. 25:51-64.

DEAN, H.S. 1953. Protection through Quarantine. U.S. Dep. Agric. Yearb. Agric. Plant Disease. 162-164. U.S. Govt. Printing Office. Washington, D.C.

HORSFALL, J.G., and A.E. DIMOND. 1959-1960. Plant Pathology. An Advanced Treatise. 3 volumes. Academic Press, New York.

HUBER, D.M., H.L. WARREN, D.W. NELSON, and C.Y. TSAI. 1977. Nitrification inhibitors—new tools for food production. BioScience 27(8) 523-529.

HUBER, D.M., and R.D. WATSON. 1974. Nitrogen form and plant disease. Ann. Rev. Plant Pathol. 12: 139-165.

KEHR, A.E. 1966. Status and opportunities for the control of nematodes by plant breeding. U.S. Dep. Agric. ARS Publ. *33-110.* Washington, D.C.

KENAGA, C.B. 1974. Principles of Plant Pathology. Second edition. Balt Publishers, Lafayette, Ind.

NELSON, R.R. *et al.* 1973. Breeding Plants for Disease Resistance. University Press, University Park, Pa.

RAINWATER, H.I., and C.A. SMITH. 1966. Protecting Our Foods. *In* Quarantines—First Line of Defense. U.S. Dep. Agric. Yearb. Agric. p. 216-224.

STAKMAN, E.C., and J.G. HARRAR. 1957. Principles of Plant Pathology. The Ronald Press Co., New York.

WALKER, J.C. 1969. Plant Pathology. Third edition. McGraw-Hill Book Co., New York.

3

Principles of Plant Disease Control with Chemicals

The safe and effective use of fungicides for the control of plant diseases is greatly influenced by a knowledge of the nature of the plant diseases for the control of which they are intended. It is also necessary to understand that there are limitations for these chemicals for plant protection. There is much more involved than merely blanketing disease-susceptible plants with fungicides selected at random. A great deal of the opposition to the use of chemicals for plant disease control is based on a widely held belief that they are all toxic to animals and humans, and are major contributors to environmental pollution. This is an unfounded and unfortunate fallacy based more on fiction than on fact.

There are no miracle fungicides that will CURE plant diseases once they are permitted to become established. Chemicals for insect control—the insecticides—are, with few exceptions, of no value for the control of plant diseases.

It is true that fungicides MAY BE TOXIC to humans and animals, but so are many products commonly used in every day living. Table salt, nail polish, aspirin, lighter fluid, hair dyes, dry cleaning products, and paint removers can be toxic when used carelessly or improperly. According to recent media publicity, even hamburgers grilled over charcoal may be responsible for human cancer. One is justified in wondering if there is anything that has not been incriminated as carcinogenic. By education we have learned to use these products with safety and they are accepted as necessary aids to present standards of modern living. Fungicides, the chemicals for plant disease control, by the same token, can be used without hazard to children, family pets, and adults if we recognize they MAY BE HAZARDOUS, like table salt, if used improperly without a knowledge of their limitations and prescribed methods of use.

Unfortunately, many fungicides have been arbitrarily grouped with parathion, DDT, mercury compounds, arsenicals, and other toxic products, and have been condemned without scientific trial as "carcinogens" capable of causing cancer in humans, and as unnecessary environmental contaminants. Nothing could be further from the truth. The fungicides currently suggested for the control of plant diseases are of a low order of mammalian toxicity, are biodegradable and are subject to extreme federal and state legal requirements before they can be placed on the market for general use. The toxicity of fungicides is discussed in detail in Chapter 11.

Many of the failures of fungicides to effectively prevent plant diseases are directly attributable to improper selection of effective products, improper application, poor timing, and failure to renew the chemical protection they provide during critical periods of pathogen infection. For every plant disease amenable to chemical control there is a preferred chemical for that particular problem. Buckshot methods of using fungicides applied on weekends or holidays, or at other convenient times, generally will not prevent the plant diseases for which they are intended. With increasing emphasis on environmental protection it is extremely important to select and use only those fungicides approved by regulatory agencies on specific crops, in order to avoid violation of regulations governing these products.

The general principles for using fungicides for the prevention of destructive plant diseases are illustrated in Fig. 3.1.

The majority of modern fungicides are used as "protectants," designed to prevent infection of susceptible plants against disease-inciting pathogens. These may be referred to as "SURFACE PROTECTANTS" of leaf, flower, fruit, and seeds of susceptible plants. Other fungicide formulations have been developed to either inactivate seed-borne fungi and bacteria or to prevent seedling infection, and are referred to as "SEED PROTECTANTS." In general, these formulations cannot be used as leaf, flower, or fruit surface protectants. Another category of fungicides has been specifically formulated to inactivate or prevent infection by soil-borne pathogens. These are referred to as "SOIL TREATMENT PRODUCTS," and are unsuited for other uses since some of them are volatile products.

Very few fungicides will ERADICATE or CURE established plant disease infections. In a few cases certain specific fungicides can eliminate local plant disease infection, in which case they are referred to as ERADICANT FUNGICIDES. The word FUNGICIDE literally means any agent capable of killing fungi since it is derived from "caedo" meaning to kill, and "fungi," fungus pathogens or molds which are inciters of plant diseases. Most of the modern fungicides do not actually kill fungi at the

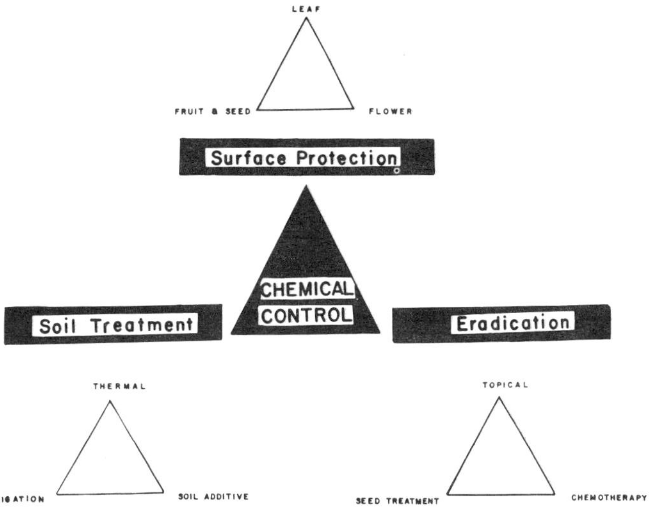

FIG. 3.1. PLANT DISEASE CONTROL WITH CHEMICALS

dosages suggested by the manufacturer. Most of them are PROTECTANTS against infections by fungi but by common usage the term FUNGUS KILLER or FUNGICIDE has become generally accepted for most of the chemicals used for plant disease control. The term PLANT or CROP PROTECTANT would be more appropriate for most of the presently used chemicals for plant disease control. In the same way the term BACTERICIDE is used for products suggested for the prevention of plant diseases of bacterial etiology. These products also are used to prevent bacterial plant diseases rather than to kill the bacteria causing the problems.

Some fungicides will inhibit or slow down a fungus pathogen without killing the organism, and these are referred to as FUNGISTATS. A fungistat will temporarily hinder the growth and development of a fungus, which, however, will renew its normal activities when removed from the influence of the fungistat. Similarly, a compound that can temporarily arrest the development of bacteria is referred to as a BACTERIASTAT, as contrasted to a GERMICIDE which destroys bacteria permanently.

Other fungicides may not prevent the growth and development of fungus hyphae and mycelium but can prevent the production of spores necessary for the initiation of secondary infections. These are referred to as ANTISPORULANTS or GENESTATIC COMPOUNDS, and have a special function in plant disease control programs.

SURFACE PROTECTANTS

As has been stated previously, by far the greatest volume of fungicides used for plant disease prevention is plant protectants applied as sprays or dusts to leaves, flowers, and fruits, prior to infection by disease-inciting pathogens. In addition, fungicidal chemicals may be applied to seeds and propagative plant parts as SEED PROTECTANTS against seed and soil-borne phytopathogens.

Most of the fungicides used as surface protectants are solids which are either mixed with inert materials and applied as dusts or treated with conditioning agents and applied as sprays (Fig. 3.2). Very few of the protective fungicides are liquids, but some have been developed for the control of specific diseases, such as dinocap (Karathane) for powdery mildew control, and Glyodin for the control of apple scab disease. Other liquid fungicide formulations are Citcop, Vapam, and Parzate, etc. A relatively recent development is the use of emulsifiable concentrates or flowable formulations such as Bravo 6/lb/gal. flowable, Captan 4 lb/gal. flowable, and Difolatan. These are cream-like formulations that may be readily mixed with water to form stable suspensions.

Most dust formulations of fungicide contain from 5–10% active ingredient incorporated with talc, infusorial earth, pyrophyllite, bentonite clay, or other inert diluents. Effective fungicidal formulations must be finely ground and of uniform particle size, and are usually manufactured commercially by the micronizing or air fractionation process. For this reason it is impractical for the amateur to attempt the preparation of dust formulations on a "do-it-yourself" basis.

One of the drawbacks to protective dust fungicide formulations is the difficulty with which adequate protective residues are maintained under unfavorable, wet weather conditions. Dry dusts applied to dry plant surfaces will be easily removed by mechanical rubbing, by blowing winds, and by heavy rainfall. For this reason fungicide dust formulations should be applied to plant surfaces wet with dew or light rain, and will usually persist better when applied in early evening or early morning when these conditions exist. Under average conditions dust formulations should be reapplied at frequent intervals in order to maintain the necessary protective residue against infection by phytopathogens.

With all surface protective fungicides it is not a question of how much fungicide formulation is applied, but how much active fungicide ingredient is deposited per acre, per 1000 square feet, per hectare, or other unit area to be protected.

As an arbitrary "guesstimate" it usually requires approximately 2 lb of active fungicide ingredient uniformly applied to one acre of growing plants for effective disease control with dusts.

Since many of the wettable powder formulations of fungicides are

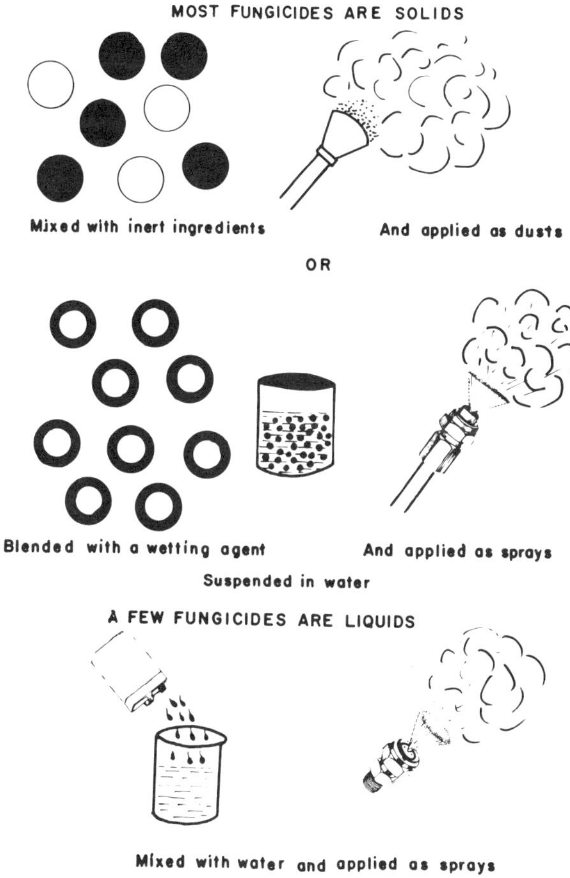

FIG. 3.2. THE NATURE OF PROTECTIVE FUNGICIDES

particulate in nature, agitation of spray preparations by mechanical or other means is important during their application. The objective of applying surface protective fungicides is to establish a protective blanket of fungitoxic or fungistatic chemical to the susceptible plant surface in order to prevent or inhibit infection by phytopathogenic organisms (Fig. 3.3).

The Importance of Correct Timing of Protective Fungicides

Correct timing of fungicides for leaf, flower, and fruit protection is the fundamental key to their successful prevention of plant diseases.

The life history of the apple scab fungus, *Venturia inequalis*, is a classical example of the necessity for correct timing of fungicides for disease prevention. This is illustrated in Fig. 3.4. *Venturia inequalis*

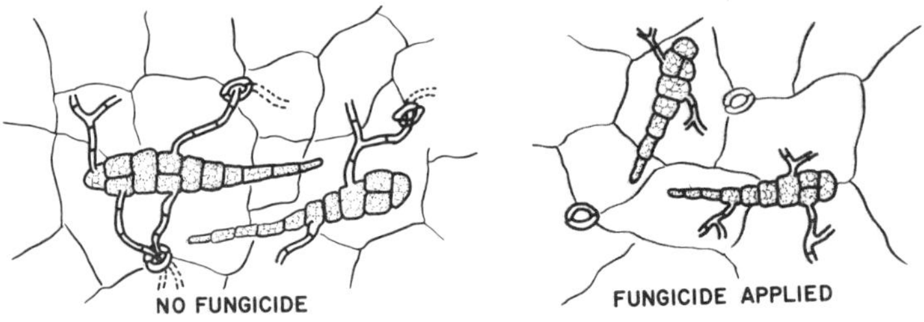

FIG. 3.3. THE MECHANICS OF FUNGICIDE PROTECTION

lives over from one season to the next in one place only—on infected leaves that fell to the ground the previous autumn. During the winter months the "apple scabber" lives on these fallen leaves infected during the previous season. It survives as mycelium and is not affected by severe winter conditions—able to withstand heavy snow cover, freezing sub-zero temperatures, and prolonged periods of dry weather.

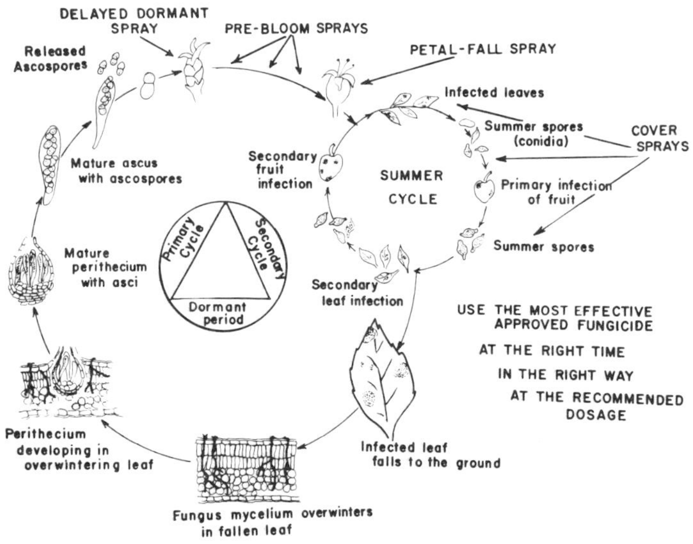

FIG. 3.4. THE TIMING OF FUNGICIDES FOR THE CONTROL OF APPLE SCAB

In the spring, with the return of favorable growing conditions, the apple scab fungus emerges from dormancy at about the same time the apple leaf buds renew their development. At this time it produces within the tissues of the infected apple leaves structures known as "PERITHE-CIA." These are destined to produce the "spring spores" (ascospores) of the apple scab fungus, the source of primary infection. By the time the apple leaf buds are ready to produce the green tips of the first developing apple leaves, the apple scabber is ready for business and produces an abundance of spring spores (ascospores) ready for discharge from the spore sacks (asci). These sacks develop within the fungus fruiting bodies or perithecia. This is a critical infection period and chemical protection must be provided to prevent "primary infections of apple scab." This is Phase 1 of chemical control of the apple scab disease. The young green apple leaf tips that protrude from opening leaf buds are very susceptible to ascospore infections. Ascospore infections should be prevented at this time, because they would be difficult to eliminate later. Apple leaf buds develop several days before fruit buds open, and failure to prevent apple scab primary infections on the first formed leaves can result in another scab year for the commercial apple grower or amateur backyard orchardist.

Primary ascospore discharge continues over a period of 4–6 weeks, and each scab spore shower causes new primary scab infections. Ascospore discharge is triggered by spring rains which moisten the overwintering ascospore-producing apple leaves, causing ascospores to be discharged like peas from a pea-shooter. This is Phase 2 of the apple scab control program, during which "delayed dormant" or "pre-bloom" protective fungicides must be applied. The number of pre-bloom fungicides will vary with the frequency of rainfall periods. In some years, when warm dry days follow the break of dormancy, a single protective spray may suffice for primary scab control. In other years with prolonged cool wet weather prior to bloom, it may require two or more pre-bloom fungicide applications.

During the in-bloom period flower bud development exposes more susceptible leaf tissue with the formation of leaves in the fruit bud cluster. At this time, in most years, there will still be ascospore discharge, and in addition, summer spores or conidia may start to develop from early primary infections. For this reason, in-bloom applications of a protective fungicide will be necessary to constitute Phase 3 of the apple scab control program.

At petal-fall time ascospore discharge should be complete, and from this point on the problem is to prevent secondary infections by conidia produced from earlier primary infections.

Phase 4 of the control program involves renewal of the protective fungicide during critical periods to prevent scab infection of developing fruits. The number of cover sprays during the summer months will depend on weather conditions, especially the frequency of periods of rainfall.

Thus, to effectively control the apple scab disease it is necessary to reapply the protective fungicide cover during critical infection periods. In most seasons, a minimum of 4–5 sprays is required in the backyard apple orchard, and 8–12 sprays in commercial apple orchards.

Correct timing of pesticide applications is equally important for the control of diseases of bacterial etiology. The example of the fireblight disease of apples and pears (*Erwinia amylovora*) serves to illustrate this situation (Fig. 3.5). The fireblight bacterium overwinters in "hold-over" cankers located on twigs and branches infected during the previous season. In the early spring, with the return of warm growing conditions, the fireblight bacteria multiply by the thousands and ooze to the surface of hold-over cankers. With the advent of spring rains and opening apple and pear blossoms, the fireblight-inciting bacteria are picked up by insects such as aphids, flies, and bees and are carried to opening blossoms. When insects contaminated with fireblight bacteria visit opening blossoms to pollinate them, the bacteria enter through natural openings, such as nectaries, and in this manner primary fireblight infections are initiated. Infected blossom spurs are destroyed, and if favorable temperature and moisture conditions occur more fireblight bacteria are produced from these sources. The secondary fireblight bacteria which ooze from infected blossoms are in turn carried by insects with sucking mouthparts to tender, succulent green terminals where they can penetrate the tissue directly, initiating secondary fireblight infections.

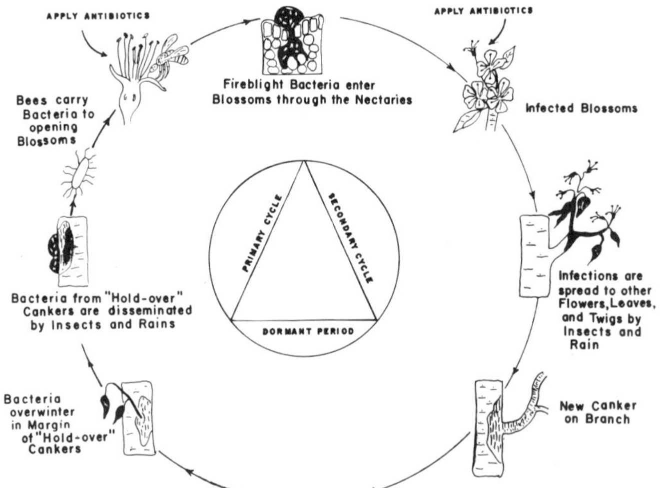

FIG. 3.5. THE TIMING OF ANTIBIOTICS FOR THE CONTROL OF FIREBLIGHT OF APPLES AND PEARS

Because fungicides are ineffective against bacteria, the use of the antibiotic streptomycin has been adopted for the control of the fireblight disease. This must be supplemented by other practices such as reduced application of nitrogen fertilizers during the blooming period, eradication of hold-over cankers, and elimination of highly susceptible apple and pear varieties.

The timing of antibiotic sprays for fireblight control is extremely critical. Application of these compounds after fireblight infections are established is useless since they do not eradicate established fireblight infections. Applications of antibiotics must be timed to coincide with primary and secondary infection periods. If this is not done they will fail to give effective control.

The use of fungicides applied as sprays or dusts for other plant diseases is also governed by the life cycle of the causal agent. There is no such thing as a universal fungicide application program for all plant diseases. The apple scab control program is more complicated than the program for other diseases such as rose blackspot, powdery mildew of ornamentals, plum and peach brown rot, and others, However, it serves to illustrate the basic principle that correct timing of fungicide applications for plant disease control is an important prerequisite for the successful use of fungicidal chemicals as partners with other practices in the "Battle of the Blights."

Further specific details on plant disease control with fungicides are outlined in Chapter 6.

The Importance of Uniform Coverage with Protective Fungicides

Unlike insect pests which move around and thus may pick up toxic residues of insecticide sprays or dusts, the phytopathogenic fungi stay put once their spores reach the infection court, and move only by hyphal growth in the vicinity of the infection site. For this reason, the principle of fungicide protection differs from the problem of insect control by protective insecticides. Fungicides, to be effective, must be uniformly applied to provide a blanket of protection against tissue penetration by fungus spores (Fig. 3.3). Since the spores of most fungi are relatively small they require only a small area of unprotected tissue to germinate and enter the susceptible host. Gaps in the protective spray deposit leave areas which can be penetrated by the hyphae or haustoria of fungus pathogens. Fungi may penetrate susceptible host tissue directly, or they may enter through natural openings such as stomata and hydathodes. Complete coverage of susceptible host tissue by protective fungicides is thus an important consideration in its use for plant disease control.

Since many fungi can penetrate the host through both leaf surfaces, fungicide coverage of both upper and lower leaf surfaces is necessary. When excessive volumes of fungicide sprays are applied, run-off of the spray droplets will result, carrying away the protective residue and leaving unprotected areas of leaves and fruits (Fig. 3.6). The most effective fungicide residues result from fine mists applied when conditions permit effective deposition of residues at the target site.

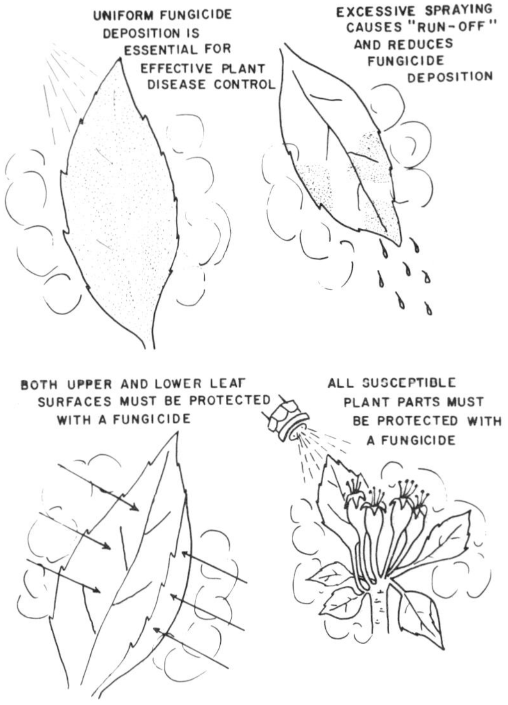

FIG. 3.6. EXCESSIVE SPRAYING REDUCES RESIDUE RETENTION

With mechanical applicators the arrangement of spray nozzles is very important for adequate spray coverage (Fig. 3.7). For the homeowner using small air-pressure sprayers, it is important to realize that the objective of disease protection with fungicide sprays is not to "soak down" the plants, but to cover leaves, fruits, and other plant tissue with a uniform residue of the fungicide being used. Many homeowners make the mistake of "drowning" the plants being sprayed with a large volume of large particle-size spray droplets. The disc in the nozzle of the small pressure sprayer is intended to break up the spray stream into finely divided, mist-like spray particles.

Most protective fungicides have a relatively short effective residual life and consequently must be reapplied to maintain residual protection of susceptible tissue. This is further influenced by the frequency of rainfall and most fungicides must be renewed at 10–14 day intervals during critical infection periods.

Another problem with protective fungicides is the rapid growth of plant

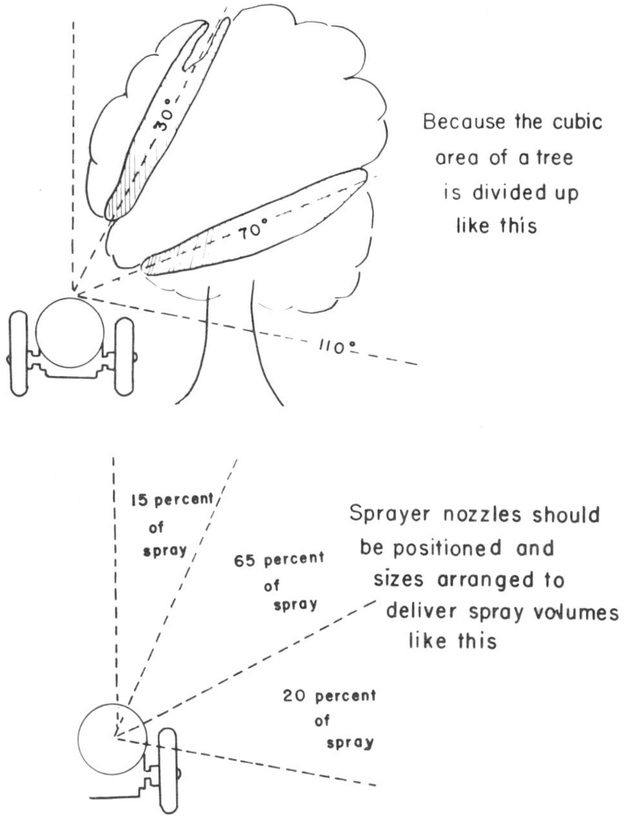

FIG. 3.7. NOZZLE ARRANGEMENT INFLUENCES SPRAY PATTERN

tissue during early spring when leaves and fruits develop rapidly. As new tissue is developed after initial spray applications it will be unprotected with fungicide. For this reason it is necessary to renew fungicide residues during periods of rapid tissue development (Fig. 3.8).

Many plant tissues are difficult to wet due to the development of waxy cuticle tissue. This is especially true with crops such as cabbage, tomato fruits, raspberry canes, and developing apple fruits. In such cases auxilliary materials such as wetting agents and sticker spreaders must be incorporated with the spray mixture to improve spray coverage and distribution. These materials are incorporated in many commercial fungicide formulations.

ERADICANT FUNGICIDES

Although most fungicides are strictly "PROTECTANTS," there are a few fungicides that are capable of ERADICATING established plant

FIG. 3.8. FUNGICIDE PROTECTION MUST BE RENEWED AS NEW PLANT TISSUE DEVELOPS

disease infections. Many of the fungicides, when used at dosages sufficient to kill fungi, will also be injurious or PHYTOCIDAL to the plant to which they are applied. For example, liquid lime-sulfur (calcium polysulfide) is an eradicant fungicide and will eliminate early infections of apple scab and peach leaf curl (*Taphrina deformans*). It is for this reason that the "old time fungicide," liquid lime-sulfur, is still included in some suggested spray schedules. Most of the powdery mildew diseases which grow superficially on infected plant surfaces may be eradicated after the symptoms appear by the use of products such as cycloheximide (Actidione) and capryl (Karathane). The fungicide known commercially as Phygon and distributed under the common name of dichlone is also an eradicant for apple scab, as well as for blossom infections of the peach and plum brown rot disease (*Monilinia fructicola*). The product known as dodine (Cyprex) also has the property of eradicating early infections of apple scab. Some of the specialized turf fungicides such as Kromad, Tersan OM, and Fore will eradicate turf diseases when applied at the recommended dosages.

In general eradicant fungicides have limited usefulness when used by knowledgeable operators, and cannot be relied upon to eradicate many of the common plant diseases. Unfortunately, there are at present no panaceas for all of the ills of growing plants.

4

Chemical Control of Plant Diseases Through the Years

Although early knowledge concerning plant diseases was clouded with erroneous concepts and superstitious beliefs, agriculturalists and philosophers of these civilizations were well aware of "blights and mildews" that affected crops upon which they depended for their daily bread. The Old Testament contains many references to blasts, blights, mildews, and smuts which evidently were accepted as visitations of an outraged Deity.

In the Homeric poems of the 8th Century B.C. the first reference to "the pest averting sulfur with its properties of devine and purifying fumigation" will be found. In the Iliad of Homer, the Greek epic poet, is described the ritual of Achilles removing a ceremonial goblet from a storage chest preparatory to a feast to Zeus, the God of Earth and Sky. In the instructions for this ceremony it was recommended that "the goblet be cleansed with sulfur and rinsed well in pure running water." The reason for this, according to the translation of "The Iliad" by Rees (1963), was that the goblet, having been stored with "tunics, windwarding cloaks, fleecy warm blankets and leather goods," was contaminated with the growth of mildew which had to be removed.

In the "Odyssey XXII," the story of the wandering of Odysseus, the husband of Penelope and the father of Telemachus, and the regaining of his kingdom is related. In this account the slaying of the suitors of his faithful wife Penelope is described. Following the slaying Odysseus called for "sulfur for purifying and fire to disinfect the Halls, House and Courtyard." It is interesting that the words "fumigate" and "disinfect" were used in this quotation from Homer.

According to the publication by the Texas Gulf Sulfur Company, "Sulfur, an Essential to Industry and Agriculture" (Anon. 1934), Dioscorides, considered to be the Greek Father of Medicine, recommended sulfur ointments for the treatment of human skin diseases which were probably of fungus etiology. In 470 B.C., Democritus (460-370 B.C.), the Greek philosopher who believed the world was made up of tiny particles (atoms) imperceptible to the senses but indivisible and indestructible, recommended "sprinkling plants with pure amurca of olives, without salt, to prevent blight from attacking them and to destroy worms adhering to their roots." This is probably the earliest recorded reference to the use of a chemical for plant protection. Marcus Portius Cato the Elder, also known as Cato the Censor, was a wealthy Roman statesman who insisted that Carthage was a city of iniquity and must be destroyed, but also published a treatise on farming. In this early handbook of agriculture he mentioned that "one way of destroying the vine fretter is to fumigate the trees with smoke from a mixture of bitumen and sulfur boiled where the smoke will blow through the trees." These examples of early attempts to prescribe controls for blights and pests of plants by chemical means in the days of the early Greek civilization are significant since they connote awareness of plant problems and the need to control them.

In the Roman civilization, agriculture was considered to be a basic national industry, and consequently it attained a high level of development. Some of the basic principles of animal and crops husbandry were initiated during the halcyon days of the Roman Empire. It is, therefore, not surprising that Roman agriculturalists interested in crop husbandry as their vocation should have noticed that plants could and did become sick, even in the days of Ancient Rome. For students of biology and others interested in plant health, it is fortunate that many of these early Roman observations were recorded for posterity and have survived in literary archives. A classical encyclopedia of Roman agriculture is the work of Caius Plinius Secundus, Pliny the Elder, entitled "Historia Naturalis." It was comprised of 37 books and a prodigious collection of second-hand information on agricultural subjects. In this early agricultural work were recorded many references to diseases of trees, grains, and grapes together with suggestions for their control. Discussing "mildew of corn," a possible reference to wheat smut, Pliny stated:

> Mildew of grain is especially prevalent when the corn is planted in a valley where there is not a good draft of wind. The steeping of the seed for planting in wine, or mixing bruised cypress leaves with it is to be recommended.

This is undoubtedly the first recorded recommendation for seed treatment for plant disease control (circa A.D. 23-79). While some of the references to plant diseases in Historia Naturalis are logical they were also mixed with superstitious beliefs and ceremonial ritual. For example,

Pliny, referring to repellents for birds that robbed the grain fields, recommended burying an unidentified herb in the corners of the grain fields, and stated:

> It is wonderful to relate, but in such a case not a single bird will enter the field. As for mildew, the greatest curse of all corn, if branches of laurel are fixed in the ground it will pass away from the field and only on to the leaves of the laurel. (*This is a possible reference to infection of barberry by stem rust.*) Many people for the more effective protection of millet recommend that a bramble frog (torch) be carried at night through the field before hoeing is done, and should be buried in an earthen vessel in the middle of the field. If this is done they say that neither sparrows, or worms will attack the crop. The frog (torch) must be disinterred before the millet is cut, for if this is neglected the produce will be bitter.

Another example of attempts by Roman agriculturalists to control a plant disease was "The Feast of the Robigalia." Evidently in the grain fields of ancient Rome rust was so destructive that every year a special ceremony was arranged to propitiate the Rust Gods, "Robigus and Rubigo." According to this legend a farmer of Rome was bothered by a red fox which nightly killed some of his prized chickens. The fox eventually was trapped alive, and the farmer's son wrapped its bushy tail with straw, set the straw afire, and released the poor fox. Everywhere the poor critter ran to escape its hot tail rust disease appeared on the grain, according to the legend. Thus, the people of Rome were considered to be responsible for epidemics of rust of cereals. The feast of the Robigalia was an annual event initiated by Numa, the second King of the Romans, to propitiate the Rust Gods for the cruelty of a small Roman boy to a marauding red fox.

Man has always been dependent on plants for food, clothing, and, in large measure, for shelter. Even though man in primitive ages ate largely the flesh of animals and clothed himself in the skin of animals, such food and clothing was possible only because of plant life on the earth.

Thus, in the ages before man learned the art of domestication and crop husbandry, diseases of the wild plants upon which he depended must have caused frequent inconvenience, perhaps famine, and doubtless at times were contributing factors to his migrations, with their attendent wars and other hardships.

The significant foundation stones upon which modern knowledge of fungicides has been erected were the early attempts in Greek and Roman agriculture to describe plant diseases and to prescribe methods for their prevention.

CHEMICAL CONTROL OF PLANT DISEASES (476–1853 A.D.)

From the fall of the Roman Empire in 476 A.D. until the 13th Century occurred the period in man's history known as the "Dark Ages," when science and learning slumbered and man was more interested in saving his soul than in protecting his crops. For plant pathologists this period could be referred to as the "Pre-DeBary Period." Throughout all of these years nothing was added to our knowledge concerning plant diseases, and evidently no attempts were made to prevent the losses they caused.

With the beginning of the 16th Century there was a quickening of man's interest in the phenomena of nature and a tendency to depart from medieval scholasticism with its dependency on the teachings of early philosophers.

The agricultural and horticultural literature of the 17th Century contains several references to plant diseases which were still imperfectly understood with little knowledge concerning their cause. Practical gardeners such as Coler in 1600, Lauremburg in 1631, and Hesse in 1690, all three of whom were probably the equivalent of practical horticulturalists, described various disorders of horticultural plants. These writings were still dominated by the thinking of early philosophers, so plant diseases were considered to be caused by superfluous saps, the stars, foul mists rising from the rheumy swamps, and other strange beliefs. Nevertheless, suggestions for the prevention of these problems were offered, which, in the light of later knowledge, were ineffective and without factual basis. However, they did indicate growing concern over the economic importance of plant diseases and the need for methods for their prevention.

One of the earliest recommendations for the treatment of tree cankers is described in a book entitled "Paradisus. The Ordering of the Orchard," published in England in 1629 by John Parkinson. In Chapter 8, page 550, Parkinson wrote as follows:

> The canker is a shrewd disease when it happeneth to a tree for it will eate the bark around and so kill the heart in a little space. It must be looked into before it hath runne too farre; most men do wholly cut away as much as is fretted with the canker, and then dress it with vinegar, or cow's pisse, or sow's dung and urine etc. until it be destroyed, and again with your salve before appointed.

A review of horticultural literature during the 18th Century reveals little or no logical understanding of effective means of plant disease control. Nevertheless, during this period plant diseases attracted the attention of plant lovers, and the activities of the plant taxonomists under the stimulation of Carl Linnaeus naturally led to observations of abnormalities in certain plant specimens. This revived interest in plant diseases. Although most of these observations were incidental to taxonomic classification of plants, methods for the control of plant diseases and plant pests continued to attract the attention of horticulturalists and practical gardeners.

Recommendations for wood preservation date back to Greek and Roman days when "astringent potions of oil extracted from olives, and oils from larch, cedar and juniper" were suggested as wood preservatives in the writings of Cato and Pliny. The first specific chemical to be suggested for prevention of wood deterioration was made in 1705 by Homberg, who, according to Hunt and Garratt (1938), recommended mercuric chloride as a wood preservative.

According to Woolman and Humphrey (1924), Jethro Tull in England reported in the London Times in 1733 that as early as 1650 Remnant had suggested the use of salt brine for steeping of wheat seed for the prevention of stinking smut. Remnant's suggestion of steeping wheat seed in salt brine evidently was the first recorded practical suggestion for chemical seed protection and was based on the observation that wheat seed salvaged from a foundered sailing vessel in the Bristol Channel, being unfit for human consumption, was used for planting purposes. Subsequently, it was noticed that the wheat seed soaked in salt water was free from stinking smut disease while adjacent fields planted with untreated wheat seed were heavily infected with stinking smut. This observation led to the practice of treating wheat seed with salt brine, followed for many years by English farmers. Jethro Tull contended that the salt brine treatment was ineffective for stinking smut control and came to the conclusion that the wheat seed salvaged from the foundered sailing vessel in the Bristol channel was not infected with stinking smut. Nevertheless, Remnant's salt brine wheat seed treatment was the first instance that a chemical was suggested as a seed protectant.

According to Woolman and Humphrey (1924), Aucanté in 1755 suggested treating wheat seed for stinking smut control with a mixture of lime, arsenic, and corrosive sublimate (mercuric chloride), which Bryant in 1784 maintained was ineffective for smut control. In 1755 Mathieu Tillet published his classical treatise on the cause of stinking smut disease of wheat. Tillet, one time Director of the Mint in Troyes, France, was not an agronomist, nor was he indeed a botanist. He was, however, a gifted investigator. His classical "Dissertation on the Cause of the Corruption and Smutting of the Kernels of Wheat in the Head and on the Means of Preventing These Untoward Circumstances" was awarded the prize of the Royal Academy of Literature, Science and Arts of Bordeaux in 1755, and was published with the authorization of the King by the widow of Pierre Brun, Licensed Printer of the Royal Academy in Bordeaux, France. Dr. Harry B. Humphrey, translator of Tillet's treatise for Phytopathological Classic No. 8, made the following statement regarding Tillet's contribution:

Tillet did not succeed in discovering the fact, later disclosed by Prévost, that the primary cause of bunt was, in truth a parasitic fungus; but he did very definitely prove that the disease resulted through contamination of the seed with the bunt dust. At no time did he suspect the true origin and function of the dust, but concluded that it was the active agent in transmitting a virus to the wheat plant, a poisonous principle that was self perpetuating in its power to reproduce more of the same sort of virus-transmitting dust.

It is truly remarkable that here was a man who, without the certain advantage of training in agronomy and botany, had the genius to carry to successful issue research of such far-reaching importance. Without instruction and alone, he planned comprehensive experimental research on an obscure problem, and it is small wonder that with his lack of knowledge of parasitism he failed to identify bunt with a parasitic fungus. He did, however, definitely prove not only the value, but the necessity, of the experimental method in research, and established forever the fact of the infective principle of bunt dust. Richly did he deserve the great honor awarded him by the Academy of Science of Bordeaux.

In 1761 Schulthess first suggested the use of copper sulfate as a wheat seed treatment. Some authorities credit this to Tessier in 1783.

The essential principles of tree surgery, including the treatment of cankers, were established by Riedel and Forsyth in 1791. William Forsyth, a gardener in the service of the King of England, published a treatise on the culture and management of tree fruits in 1802. In this publication he outlined directions for the preparation of a concoction "for the cure of diseases, defects, and injuries of trees, being particularly valuable in the healing of wounds." Forsyth's treatment for tree cankers became widely used for filling cavities in trees. His wound dressing mixture was prepared as follows:

Take one bushel of fresh cow dung, one half bushel of lime, using rubbish from old buildings, one half bushel of wood ashes, one sixteenth bushel of pit or river sand. The last three are to be sifted fine before they are mixed. Then work them well together with a spade and afterward with a wooden beater until the mixture is very smooth, like fine plaster used for the ceiling of rooms. Soap suds are to be recommended to make the composition the consistency of paint. After this mixture is applied it should be covered with a sifting of powder of wood ashes mixed with a sixth part of the same quantity of plaster or paint.

PLANT DISEASE CONTROL IN THE 19TH CENTURY

Sulfur undoubtedly is the oldest of all pesticides. It was recommended by Forsyth in 1802 for the control of powdery mildew on fruit trees. He recommended a mixture of quick lime, sulfur, elderberry juice, and tobacco. In 1824 Robertson reported at the meeting of the London Horticultural Society that sulfur was the only effective remedy that he knew of for the control of mildew on peaches, and recommended the addition of soap to improve the effectiveness of the sulfur spray. Robertson was thus one of the first to suggest the use of a surfactant to improve pesticidal properties. He also emphasized the importance of complete coverage for disease control. A few years later, in 1841, Mearns recommended a mixture of alcohol and sulfur for the control of powdery mildew on peaches. He recognized the importance of early applications and suggested that the mixture be painted on the branches during dormancy in early spring.

In 1807 Prévost published his classical, "Memoir on the Immediate Cause of Bunt or Smut of Wheat, and of Several Other Diseases of Plants, and on Preventives of Bunt." In this publication, translated by Dr. George W. Keitt in 1939, Prévost reported on critical studies of the effect of different concentrations of copper sulfate, temperature, and duration of treatments on the germination of wheat bunt fungus chlamydospores. Horsfall (1945) considered the work of Prévost as a pioneer example of laboratory testing of fungicides.

The first suggested use of self-boiled lime sulfur was made by Kendrick in the U.S.A. in 1833. In 1834 Knight recommended sprinkling peach trees with sulfur and lime for the control of peach leaf curl (see Pierce 1900).

Around 1850 the powdery mildew disease of grapes (*Uncinula necator* (Schw. Burr.) evidently was introduced into England on grape cuttings from the early pioneer vineyards of America. While the fungus etiology of this disease had not been proven at that time, the British mycologist Miles Joseph Berkeley, a minister of the church, suspected its true nature and described the fungus associated with the disease.

The 1845 outbreak of the destructive powdery mildew disease of grapes in England has been described by Large (1940) as follows. The reading of this text, entitled "The Advance of the Fungi," is recommended as a most interesting account of the early history of several destructive plant diseases.

The grape vines near Margate, a seaside resort in England, had been attacked since 1845 by a peculiar malady, never observed before. Every vine, whether in the open borders or in (green) houses, was affected. The disease appeared on the young shoots, tendrils, and leaves, like a dusting of white and pulverulent meal; it spread rapidly on to the grapes themselves, withering the bunches when they were small and green, or causing the grapes to crack and expose their seeds when they were attacked later. The disease was accompanied by an unpleasant mouldy smell, and it ended in the total decay of the fruit. But one, Mr. Tucker, gardener to J. Slater Esq. of Margate, had not stood by, wringing his hands and saying it was the will of the Almighty that the murrain on the potatoes should now spread to the vines. On the contrary, and in the exercise of an initiative most exemplary in a labouring man, he had obtained access to a powerful microscope and examined the afflicted foliage. He saw that a powdery mildew was growing on it, and then without waiting for the fungus to be identified, and untroubled by any philosophic doubts as to whether it was the cause or consequence of the disease, he looked about him for a practical way of stopping it. For once in the history of plant pathology, the announcement of the outbreak of a new disease was coincident with the publication of an effective remedy for it.

Mr. Tucker was accustomed to using sulfur as a specific against mildew on peach trees, and he now tried sulfur for this new trouble on the grape vines. He made up a mixture of flowers of sulfur and lime in cold water and applied it to the leaves with a sponge. The treatment worked admirably: the mildew was killed on the affected shoots, its spread was prevented, and the rest of the grapes and the foliage grew away unharmed. In the end there was a good crop in the houses under Mr. Tucker's care, while in some of the other vineyards in the locality there was not a single sound bunch of grapes to be seen.

Mr. Tucker, signing himself "Progressionist," published a humble account of his observations in a letter dated September 22nd, 1847, to the popular Gardener's Journal, and the Rev. M.J. Berkeley, in the more influential Gardener's Chronicle, pronounced that the causal fungus was a species of "*Oidium*" which he named *Oidium Tuckeri*. Mr. Tucker's name was destined not only for immortalization in the dusty records of cryptogamic botany, but for immediate, widespread and mournful celebrity.

So successful was Mr. Tucker's recommendation for the control of grape powdery mildew, using brimstone applied to the grape leaves with a sponge, that it became for several years a generally adopted practice by English peach growers.

Shortly after the outbreak of peach powdery mildew in England it skipped across the English Channel to cause extensive damage in commercial vineyards in France. Professor Duchartre, of the Institute Agronomique at Versailles, was one of the first to use Mr. Tucker's sulfur remedy in French vineyards. While the treatment proved to be effective and kept the Oidium in check, sulfuring grape vines over hundreds of hectares of open hillside proved to be time-consuming, arduous, and expensive. Consequently, several labor-saving schemes were suggested and tried. One suggestion was to wet the grape leaves with plain water first, and then to blow on dry sulfur dust with specially constructed bellows. A Count Duchâtel used this method successfully on 1,200 acres of his grape vines. The treatment was quite effective but it was soon found that preliminary wetting of the vines could be omitted if the sulfur dust was applied when the grape leaves were wet with dew. Frequently a perfunctory dusting of the grapevines with sulfur, regardless of whether the grape leaves were wet or dry, appeared to be equally effective. Thus it was that the practice of dusting with a fungicide for the control of a destructive plant disease was used for the first time in France.

In 1851 the head gardener at the Royal Palace of Versailles, France, was confronted with the urgent need to control a devastating outbreak of grape powdery mildew. He suggested using a mixture prepared by boiling equal parts of ground brimstone with freshly slaked lime for ten minutes. This mixture became known as "Eau Grison," after its originator, and was widely used as a practical control for surface mildews.

While amateur gardeners and others were using "buckshot" methods for seeking out chemicals capable of preventing plant diseases, events were rapidly moving towards biological history that was destined to dispel for all time the theory of spontaneous generation (development of lifeless matter into living forms). In 1822, in the French village of Dôle, Louis Pasteur was born, the son of a French tanner. He attended the École Normale in Paris where he studied chemistry, and later entered the Sorbonne to continue his studies in chemistry. In 1867 he became Professor of Chemistry at that institution, and it was during this tenure that he began studies of living matter from which were to come contributions that were destined to be invaluable to the health and well-being of mankind. Some of his earliest experiments with alcoholic liquors that had "gone sour" were preliminary to his later research that eventually dispelled the myth of spontaneous generation. With the aid of a microscope, Pasteur discovered that germs had entered the "diseased liquor," and that heating the liquor destroyed the germs. This discovery is said to have saved the French Government more money than was required as indemnity following the Franco-German war.

When the silkworm farmers of southern France found their silkworms dying and the mulberry trees deserted, they appealed to Pasteur to cure the disease that threatened their livelihood. Pasteur, after a number of tests, found that the little specks on the silkworms were caused by bacterial parasites, and he quickly showed the French farmers how to combat them successfully. The silk harvest returned to normal and

Pasteur was acclaimed for having saved the French silk industry.

Later, Pasteur worked with animal diseases such as cholera in chickens, and anthrax in sheep, proving they were both caused by bacteria. Through his efforts to cure animals of these diseases, he worked out the principles of inoculation whereby he introduced into the blood stream of healthy animals weakened germs of the deadly diseases in the form of "vaccine." Animals treated in this manner were able to resist the disease against which they had been vaccinated. His later work on hydrophobia and rabies added more evidence in favor of the germ theory of disease. Pasteur, opposing the theory of spontaneous generation, had declared, "Gentlemen, I regret the necessities of your detestable philosophies, but spontaneous generation of small organisms does not occur in my flasks." DeBary, in his classical paper, "Development of Some Parasitic Fungi," published in the Annales des Sciences Naturelles in 1863, supported the claims of Pasteur in his statement, "Gentlemen, I have the honor to advise you that it does not occur among my little fungi either, and as this would seem to cover the whole of your present ground, we may now adjourn. Omne vivum e vivo. Good Day."

In 1853 Heinrich Anton DeBary published his classical contribution on the causative role of fungi as incitors of plant diseases, and thus became the Father of Plant Pathology. DeBary was greatly influenced by the discoveries and teachings of Pasteur and these were the biological foundation on which he built his contribution to the true nature of plant diseases.

Following DeBary's proof of the true nature of plant diseases, developments in chemical controls for plant diseases occurred rapidly. "Danish Hot Water" was discovered and introduced by a Danish schoolmaster named Jens Ludwig Jensen, who was the editor and distributor of an agricultural magazine during the period 1868-1880. In 1881 he organized his own private agricultural research institute which he called "Bureau Ceres," located in the city of Copenhagen. Jensen was aware of the serious nature of the late blight disease of potatoes and published privately a pamphlet entitled, "How to Overcome the Potato Disease." In this publication he emphasized the need to eliminate the source from which the disease originated, and asked the question, "What was there then, under Heaven, that could be put into the potato to kill the tangle of fungal threads interwoven among the cells, without harming the potato itself?" He suggested the answer to the riddle was heat, and thus pioneered thermal sterilization for the elimination of phytopathogens. In 1887 Jensen published his method for the eradication of *Phytophthora infestans* (potato late blight fungus) in the Mémoirs de la Société Nationale d' Agriculture of France and received the Society's Gold Medal for his contribution. In this publication he suggested the use of dry heat, exposing infected potato tubers to a temperature of 40°C for 4 hours, and illustrated a simple apparatus to expedite the treatment.

It is not generally known that the "hot-water treatment" for loose smut of wheat FOLLOWED the demonstration by Jensen of the practicability of thermal treatment for the eradication of the potato late blight fungus. Jensen, working in his private research institute in Copenhagen, discovered that there were two smuts of barley: "loose smut" and "covered smut." He confirmed the suspicion that loose smut of oats did not affect barley or wheat, and indeed demonstrated that there were three fungi responsible for these wheat smut problems. Previously it had been considered that one species of smut was responsible. Jensen discovered that not only did the hot water treatment kill the spawn of smut fungi within cereal seeds, but it also killed smut spores between the kernel and the husk, and on the surface of the outer seed coat as well. It not only prevented loose smut, for which no remedy had previously been found, but was also as effective as copper sulfate against the stinking smut of wheat caused by *Tilletia tritici*. Jensen discovered all of the smut diseases of cereals, loose and covered, could be prevented by soaking the seed for half a day in warm water (68−86°F), and then plunging it for 5 min into hot water between the precise limits of 127−137°F, depending on the type of grain. Jensen's hot-water treatment for cereal smut control was adopted widely in Denmark, and even penetrated the conservative Academic Circles of Great Britain. Jensen's classical article, "The Propagation and Prevention of Smuts in Oats and Barley," was published in the Royal Agricultural Society Journal in 1888. It was in the U.S.A. that Jensen's hot-water treatment was thoroughly investigated and then introduced into general use by W. T. Swingle in 1892. Jensen's contributions thus pioneered the use of thermal sterilization for the elimination of phytopathogens causing certain diseases of some economic crops.

The dramatic discovery of bordeaux mixture for the control of grape downy mildew disease (*Plasmopara viticola*) has been described by some authors as equal in biological significance to Pasteur's classical research on the germplasm theory of disease. Millardet's work with bordeaux mixture must at least be considered as a classical landmark in the history of chemical plant disease control, and the beacon that lighted the path for the rapid progress that was to follow.

Toward the end of October, 1882, Pierre Alexis Millardet, a professional entomologist, was strolling through the vineyards of Saint-Julien in Médoc, in southern France. Downy mildew had been a destructive problem that year, and Millardet was surprised to notice that the vines beside the path were still in leaf, while elsewhere grapevines were bare and devoid of fruit and foliage. This so aroused his curiosity that he inquired from Monsieur Ernest David, manager of the vineyards at Château Beaucaillon, as to what had been done to the mildew-free grapevines. He learned that it was the custom of the grape growers of the Médoc to treat the vines along the path with a conspicuous and poisonous-looking substance to discourage small boys and passersby from stealing the grapes. Monsieur David informed Millardet that a mixture of copper sulfate and lime was used for this purpose. Further observation showed that indeed all of the vines along the path had been "Pilferproofed" and were greener and free from downy mildew. Millardet was greatly excited and made plans for precise field experiments the following year.

In 1883 Millardet tried verdigris, bluestone alone and mixed with lime, and salts of calcium and iron as preventatives of downy mildew. Alas, 1883 was a dry year and very little mildew developed in the experimental field demonstration plots. But Millardet reported that the experimental treatments "did some good," especially the bluestone or copper sulfate treatment. The following year Millardet set out extensive field plots for mildew prevention. His treatments included carbonate, acetate, sulfate, phosphate, and sulfide of copper, with and without lime, lime alone and the green vitriol or sulfate of iron. However, it was another year with little mildew. Nevertheless, Millardet published his findings in the May, 1885, Proceedings of the Society of Agriculture for La Gironde. His precious formula for mildew control was as follows:

> Dissolve 8 kilos of copper sulfate crystals in 100 liters of water. In a separate vessel slake 15 kilos of good quicklime and stir up with 30 liters of water. The "milk of lime" should be poured into the copper sulfate solution with constant stirring to produce a bluish paste.

This was the first bordeaux mixture, or "Bouillie Bordelaise." In 1885, the owner of the vineyards at Dauzec and at the Château Beaucaillon treated 150,000 grapevines with "Millardet's Mixture" for downy mildew control. The year 1885 was a plant pathologist's dream—ideal wet weather prevailed and downy mildew was severe. The field demonstrations of "Millardet's Mixture" came through with flying colors. In his report, Millardet wrote:

> On October 3rd the leaves of the treated vines at Dauzec were healthy and of a beautiful green, the grapes black and perfectly ripened. The rows of vines left between them for comparison were in a wretched state; the majority of their leaves had fallen, the few remaining were dry and shrivelled, and the grapes were

still red, unfit for anything but sour wine. The contrast could not have been more striking.

The same spectacular results were obtained with bordeaux mixture not only in other vineyards of Médoc, but as far away as Italy, where M. le Marquis Pinelli, of the Château Tagliolo, had treated 80 hectares.

Some people have said that the dramatic success of bordeaux mixture for the control of grape downy mildew could be compared with Louis Pasteur's successful control of sheep anthrax on the farm of Pouilly le Fort. E.C. Large, in his textbook, "The Advance of the Fungi," describes in detail the controversy that followed the success of Millardet's Mixture, as to who was the first with the answer for the control of the devastating mildew disease of grapes.

Large (1940) describes how there were some who said it was not Millardet who discovered Bouillie Bordelaise, "but that God, working in his mysterious way, had revealed it by the miracle of an accident to his children of Bordeaux as a reward for their sweet and charitable disposition, perhaps in only pretending to poison the grapes that might tempt the passing stranger."

Millardet, writing in the Journal d'Agriculture Pratique in the fall of 1885, eulogized his own achievements in the style of an obituary notice:

> I claim the honor of having been the first to conceive the idea of the treatment with copper: the first to experiment with it, and the first also to recommend its use in practice. And may I be permitted to add—for these things are for us savants our titles and our most precious trophies—that in 1878, simultaneously with Monsieur Planchon, it was I who first observed the presence of mildew in France, and that since then I have been constantly on the alert. To this my work bears witness.

Millardet's critics accused him of promoting a chemical mixture already known for plant disease control, and in engaging in unseemly scrambling for professional credit. Despite this, on his death in 1902 the grateful people of Bordeaux erected a monument in his memory. On the monument, which consists of a bust of Millardet on a grape sculptured granite plinth, the following inscription is engraved:

> P.M. Alexis Millardet. Professor of the Faculty of Sciences of the University of Bordeaux. Correspondent of the Institute of France 1838–1902. In memory of his work on the maladies of the vine.

The work of Millardet and his associates must be recognized as the first practical investigation of chemicals for the control of a destructive plant disease—the beacon that lit the path of rapid progress that was to follow. By the end of the 19th Century, fungicides, the chemical tools for plant disease prevention, while few in number, had been accepted as supplementary tools in the Battle of the Blights. Man had advanced from superstitious beliefs and groping in the dark to a logical and scientific understanding of biological phenomena, and application of logical methods for prevention of factors limiting crop production, such as plant diseases.

FUNGICIDES IN THE 20TH CENTURY

At the start of the 20th Century, bordeaux mixture and lime-sulfur reigned as undisputed kings of the fungicide world. Both of these compounds were universally used for the control of all plant diseases. It was soon discovered, that while these materials had their virtues, they also had their faults. Both lime-sulfur and bordeaux mixture were excellent fungicides, but neither was pleasant to use, and both could be injurious to foliage and fruit under certain environmental conditions. In the U.S.A. fruit growers and other agriculturalists interested in the use of fungicides began to demand "safer chemicals" for plant disease control.

Following Millardet's introduction of "Bouillie Bordelaise," substitute formulations of copper were suggested for plant protection. Eau Celeste, consisting of ammonia as a substitute for lime in combination with copper sulfate, was one of these introduced by Audoynaud in 1886. It proved to be injurious to most foliage and never competed with bordeaux mixture. Another substitute was Burgundy Mixture, or Soda Bordeaux, introduced in 1887 and consisting of sodium carbonate as a substitute for lime in combination with copper sulfate. This also proved to be inferior to bordeaux mixture.

The introduction of self-boiled lime-sulfur by W.M. Scott in 1908 offered temporarily a less injurious fungicide for fruit disease control than liquid lime-sulfur or bordeaux mixture. Scott's self-boiled lime-sulfur was prepared by slaking together 15 lb of quicklime and 10 lb of finely divided sulfur in a small amount of water with constant stirring. As soon as yellow streaks of polysulfide began to appear, the mixture was quenched with water. The entire batch was further diluted with water to make 50 gal. of spray. Obviously, self-boiled lime-sulfur was messy to make and difficult and unpleasant to apply, but it was used for several years by peach growers in the U.S.A. as a less injurious substitute for liquid lime-sulfur.

By 1907 faint glimmerings of interest in some reliable technique for fungicidal bioassay was becoming evident primarily as a result of the work by a German forest pathologist named Falck. He was interested in the preservation of wood and demonstrated that the rate of growth of a fungus on solid cultural media was linear with time, and that the presence of a toxicant in the medium did not alter this relationship. It was this contribution that provided one of the standard methods now widely used in the quantitative study of fungitoxicity. Many of the fungicides introduced in the mid-1900's were spawned by laboratory "in vitro" bioassay.

In 1913, following the publication of Ehrlich's research on the value of organic arsenic for the treatment of human syphilis, Riehm introduced the organic mercuries as seed treatment chemicals for the control of stinking smut of wheat. The sponsoring industry for the organic mercurials was I.G. Farben Industrie in Germany. This was soon followed by the introduction of several mercury seed treatment chemicals by the DuPont Company of Wilmington, Delaware.

In 1917, an Australian worker, Darnell-Smith, introduced copper carbonate as a dust seed treatment for wheat, but this was short-lived due to its unfortunate property of generating so much friction in the drill box that excessive mechanical damage resulted.

In 1918 the Canadian workers, Saunders and Kelsall, working in the Annapolis Valley Fruit Research Laboratory in Nova Scotia, introduced copper-lime-dust as the first fungicide in dust form to be widely used for plant disease control.

In 1920 the Crop Protection Institute was established at the University of New Hampshire. This unique agency was designed to serve as a liaison between the agricultural chemical industry and the agricultural experiment stations to promote and sponsor fundamental and applied research on more effective fungicides. Later contributions of this agency fully justified its establishment.

In 1930 McCallan published the first in a series of classical papers on the dynamics of fungicidal action. It was the research work of this careful investigator that laid the ground work for the modern laboratory and greenhouse evaluation of candidate fungicides. This was to provide in later years a quick and reasonably reliable method for fungicide evaluation, leading to the introduction and adoption of many new fungicides (McCallan 1967).

By 1932 bordeaux mixture was beginning to lose out, and the low-soluble, or fixed coppers, were being introduced, soon to be accepted as effective and safe substitutes for bordeaux mixture. By this time, the development of inorganic fungicides such as copper, sulfur, and mercury had attained its peak. The year 1932 marks the "Dawn of the Organic Fungicide Era" which led to the development of most of the fungicides in use in 1968.

THE ORGANIC FUNGICIDE ERA (1930–1960)

In 1931 the U.S. Department of Agriculture established a project to search for and to develop non-corrosive and non-phytocidal substitutes for the standard fungicides in use at that time. This project was placed under the direction of J.W. Roberts, M.C. Goldsworthy, and E.L. Green, and as a result of their joint research a better understanding of the fundamental characteristics of low-soluble copper fungicides resulted. Methods for copper analysis were developed and this made possible the determination of extremely small quantities of free copper remaining on plant leaf surfaces treated with copper fungicides. Bioassay methods also were developed to distinguish between available copper and soluble copper which permitted the more precise determination of the fungicidal potential of various fixed, or low-soluble copper compounds.

While Goldsworthy and Green were doing research to develop substitutes for bordeaux mixture as fruit fungicides, other research workers were trying to find similar substitutes as fungicides for the control of vegetable diseases. Thus, in 1932 Horsfall was working at the Geneva (New York) Agricultural Experiment Station studying the value of low-soluble copper compounds for the prevention of damping-off disease of vegetables. This later resulted in the introduction of red cuprous oxide and yellow cuprous oxide as seed protectants for vegetables. In 1936 Sessions, working at the New Jersey Agricultural Experiment Station, demonstrated the potential of copper zeolite as a fruit fungicide, which was later introduced commercially under the trade name "Coposil." Nikitin (1937), working at Columbia University, studied the fungicidal properties of copper zeolite, which later resulted in the introduction of this compound under the trade name "Z-O." As a result of these research activities, colloidal coppers, basic copper sulfate, cuprous oxides, basic copper sulfate, copper oxychloride, copper phosphate, and copper silicate soon became accepted as substitutes for bordeaux mixture for the control of fruit and vegetable diseases.

In the 1920's rubber chemists interested in accelerating the vulcanization of rubber with sulfur discovered several rubber accelerators, among which was tetramethylthiuram disulfide. Tisdale and Williams, working for the DuPont Company, found that these rubber accelerators had unusual antifungal and antibacterial properties. In 1934, U.S. Patent 1,972,961 was assigned to the E.I. DuPont de Nemours and Co. "to investigate derivatives of dithiocarbamic acid as disinfectants to control and prevent the growth of fungi and microbes." However, the carbamate fungicides were not introduced at that time, but Tisdale and Williams definitely set up the signpost directing the pathway to an entirely new category of fungicides— the dithiocarbamate fungicides. In the same year, Hubert Martin, the British fungicide research worker, discovered independently the fungicidal activity of the dithiocarbamates. It remained for Hamilton in 1941 (Hamilton and Palmiter 1943) to report for the first time the effectiveness of ferric dimethylidithiocarbamate (Ferbam) as an effective control for apple scab (*Venturia inequalis*).

In 1938, simultaneous and independent research by the Imperial Chemicals Industry of England and the U.S. Rubber Company lead to the discovery of the fungicidal properties of the quinone chemicals. Working under the direction of J.G. Horsfall at the New York (Geneva) Agricultural Experiment Station, Sharvelle and Cunningham uncovered the first organic seed protectant (Cunningham and Sharvelle 1940). This was the organic compound 2, 3, 5, 6-Tetrachloro-1,4-benzo-quinone, which was later assigned the common name "Chloranil" and was marketed as a seed protectant by the U.S. Rubber Company (later known as the Uniroyal Chemical Co.) under the trade name Spergon. Horsfall has expressed the opinion that the "introduction of the quinone fungicides accelerated the faltering industrial program on the development of the dithiocarbamate fungicides and opened a new era in plant pathology. It made organic fungicides possible."

Following the introduction of chloranil as a seed protectant, terHorst and Felix introduced 2,3-dichloro-1,4-naphthoquinone as a fungicide (terHorst and Felix 1943). This was assigned the trade name Phygon and designated by the common name "dichlone." It found wide usage as a seed protectant and organic fruit fungicide.

As a result of the "Quinone Scoop of 1943," the DuPont Company renewed their research on derivatives of dithiocarbamic acid as fungicides. In 1943 Fermate was reported as "a promising fungicide" and in 1949 was assigned the common name of "ferbam." It was accepted quickly by commercial fruit growers as an effective foliage and fruit protectant for the control of apple scab and apple cedar rust diseases. Other carbamate fungicides quickly followed. In 1941 Harrington first reported that "thiram" (tetramethylthiuramdisulfide) was effective for the control of turf diseases. However, it was not formulated and sold as a commercial fungicide until 1951. The seed protectant known as "Arasan" followed some years later.

In 1944 Heuberger introduced zineb (zinc dimethyldithiocarbamate) for the control of early blight and anthracnose of tomatoes, and early blight of potatoes. In 1943 Hester of the Rohm and Haas Company was granted a patent for nabam (disodium ethylenebisdithiocarbamate), a liquid fungicide with tenacity sold under the trade name Dithane D-14. J.W. Heuberger's discovery of the "Heuberger Effect," stabilization of nabam with zinc sulfate, led to the introduction of zineb (zinc ethylenebisdithiocarbamate) in 1949 (Heuberger and Wolfenbarger 1944).

This product has been used widely as a vegetable fungicide, distributed under the trade names Parzate and Dithane Z-78. Zineb soon was followed by maneb (manganese ethylene bisdithiocarbamate) in 1954, and finally by the maneb zinc ion-coordinated product sold commercially as Dithane M-45. This compound also has been used widely as a vegetable fungicide.

In 1946 Wellman and McCallan, working at the Boyce Thompson Institute, introduced the imidazolines, or glyoxalidines, as protective fungicides for the control of certain fruit diseases. The most interesting of these was 2-heptadecyl-2-imidazolin acetate, a fungistat introduced under the trade name Glyodin in 1953. This product was used extensively for the commercial control of apple scab and cherry leafspot disease.

In 1952 "Kittleson's Killer" was introduced as the "Miracle Fungicide" under the trade name Captan. This product, with Trichloromethyl dicarboximide as the active ingredient, has had a sensational impact as a broad spectrum fungicide for the control of fruit, vegetable, and ornamental diseases, as well as a seed protectant and soil fungicide. McCallan considers the development of Captan by Kittleson to be "the introduction of the most important group of fungicides since the advent of the dithiocarbamates." An analog of captan, introduced under the trade name Phaltan, and the assigned common name Folpet, appeared in 1961 and also has been used for the control of a wide range of plant diseases.

The guanadine fungicides were introduced in 1959 by the American Cyanamid Company under the trade name Cyprex, and proved to be a very effective protectant for the control of apple scab.

THE CARSON CRISIS

In 1962 the publication of the late Rachel Carson's "Silent Spring" really focussed attention on pesticides with a vengeance. This well intended but misguided effort of a competent marine biologist, author of "The Sea Around Us" and "The Edge of the Sea," to pose as an authority on chemical pest control, resulted in a nationwide wave of concern over the hazards of chemical pesticides. It is justifiable to point out that Rachel Carson made little or no reference to fungicidal chemicals in "Silent Spring," which was devoted almost entirely to dramatizing the environmental hazards of insecticides, with passing mention of herbicides. Unfortunately the general public did not distinguish between chemical weapons against insects (insecticides) and chemical tools for plant disease control (fungicides). Thus, Rachel Carson's "Silent Spring" was used by bird lovers, conservationists, television entertainers, and politicians to implicate fungicides as poisons on the American dining table. Long before the birth of "Silent Spring," industrial producers of fungicides and plant pathologists who advocated their use were aware of the hazards of fungicides and had made an honest effort to protect the public against possible misuse of these new chemical tools for the essential control of destructive plant diseases.

Eight years before the appearance of "Silent Spring," the U.S. Congress passed the Miller Pesticide Amendment of the Federal Food, Drug and Cosmetic Act of 1938, and this became known as Public Law 518. The Miller Amendment initially was intended to provide protection for the public against misuse and abuse of new organic chemical pesticides. However, the nationwide concern aroused by "Silent Spring" influenced legislators to make the necessary law an effective barrier to the further development of organic pesticides. As a result of this anti-pesticide movement, which was reflected even in the research programs of the agricultural experiment stations, no individual or small corporation could afford to meet the requirements imposed by new regulations governing the development introduction, and marketing of pesticides in the U.S.A. Only the larger agricultural chemical corporations could afford the costly, time-consuming, long-term feeding trials required by federal law to obtain label approval required for the introduction of a new pesticidal chemical. Torgeson (1967) estimated that the 1965 cost of developing an agricultural chemical from initial screening to development costs including field trials, development of analytical methods, residue and toxicological research, and patent costs was $2,918,000. With the inflation that has resulted since 1965, this cost in the 1980's is in the vicinity of $3.5−4.0 million. There has been little or no progress in developing new fungicides in the U.S.A. since 1968. In 1964, the Elanco Products Division of the Eli Lilly company released the powdery mildew specifics Tutane (2-amino butane) and Pipron (Propyl dichloro benzoate). Bravo (tetrachloroisophthalonitrile) was introduced in the same year by the Diamond Shamrock Company. The oxathiins were introduced by the Uniroyal Chemical Company in 1966 under the trade names Vitavax and Plantvax. Benomyl (Benlate, Tersan-1991) was introduced by DuPont in 1968. Truban, also known as Terrazole, was introduced by the Olin Mathieson Corporation in 1969. Since this time there have been no new fungicides developed in the U.S.A. and registered for use in commercial agriculture. It is of interest to note that new fungicides such as Bay-Meb 6447 were introduced in Europe in 1975 by Bayer AG of Germany, and other fungicides have been developed in Japan since 1970.

On November 28, 1975, the Federal Insecticide, Fungicide and Rodenticide Act was further amended under the provisions of Public Law 94−140. This virtually eliminated any new major developments in the area of fungicides in the immediate future. The cost to the American public for implementation of FIFRA amounted to $47,868,000 for the period October 1, 1975–September 30, 1976. A further appropriation of $23,600,000 was assigned for the EPA program for the period October 1, 1976–March 31, 1977. Under FIFRA, as amended, the EPA is responsible for registering all new pesticides. It also was supposed to review registrations either to reregister or to initiate cancellation proceedings for specific pesticide usages by October 21, 1977. A large backlog of both new registrations and reregistrations developed, however, and new registrations of pesticides have been at a virtual standstill. In order to add further confusion to the Federal Pesticide Regulations, the EPA introduced a new regulation designated as RPAR, "Rebuttable Presumption Against Registration." This regulation requires that any pesticide in which there is some evidence that unreasonable risk to man or the environment may occur must be subjected to an evaluation that includes a detailed assessment of risks and benefits.

The agricultural pesticides are not alone in their regulation by the government. The degree to which the U.S.A. has become federally regulated is well illustrated by the satirical description of regulation by Paul L. Filter, Senior Public Relations Specialist of the Dow Chemical Company, Midland, Michigan (Filter 1977):

In the morning you are awakened by the clock radio. The power for the clock came into the house from a utility regulated by the Federal Power Commission (FPC) and your State Public Service Commission. The radio station that came on was regulated by the Federal Communication Commission, with the commercials approved by the Federal Trade Commission. You threw back the bedsheets, cotton of course, raised under a U.S. Department of Agriculture subsidy and acreage allotment. The thread count of the cotton sheets was regulated by the U.S. Department of Commerce. You went into the bathroom and switched on the lights—electricity regulated by the FPC—and turned on the shower—water from the City Water Department, monitored by the Environmental Protection Agency (EPA), and heated by gas regulated by the Interstate Commerce Commission, the Federal Energy Administration and the State Public Service Commission. You applied shaving cream from a can with a child-proof closure approved by the Consumer Products Safety Commission, brushed your teeth with a toothpaste approved by the Food and Drug Administration and then sat down to breakfast—orange juice, eggs and bacon, all inspected by the Department of Agriculture.

You got into your car, inspected and approved by the State Police. Incidentally, your driver's license and automobile registration were issued by your state Department of Licensing and Motor Vehicle Bureau, and you probably drove on a highway regulated by the Federal Highway Commission.

At work you wrestled with the problems of the day—most of them regulated in one way or another—then, finally, you went home to a relaxing regulated evening.

Dr. George L. McNew, in the 50th Anniversary Edition of the American Phytopathological Society, Plant Pathology Problems and Progress. 1908–1958, concluded his discussion "Landmarks During a Century of Progress in the Use of Chemicals to Control Plant Diseases" with the following comments:

Many landmarks have been established in the realm of scientific agriculture during the past century. None of these rises more prominently on the plains of human endeavour than those that mark the forward progress in controlling plant diseases by chemicals. They stand as monuments to the research and ingenuity of many careful investigators.

The pillar of each monument that meets everyone's gaze is the tremendous economic gain from control of the fungi, bacteria and nematodes that makes it possible to feed and clothe people better than would have been considered possible when the American Phytopathological Society was first founded. Less conspicuous, but very necessary to each of these pillars, is its firm foundation consisting of sound knowledge and principles that were uncovered while the battle for crop productivity was in progress. These foundations are the hope for the future just as the pillars are the monument to the achievements of the past. From these principles will come the ideas that will guide the scientists of tomorrow in solving the problems of another generation.

The unresolved problems continue to loom large on the horizon. The farmers of the United States still endure a loss of over 7 percent of crop productivity because of plant diseases. This is a toll upon the farmland of about three billion dollars annually that may be alleviated by new types of chemicals and better methods of using those that are already available. There is little room for complacency or self-satisfaction from the discovery of new chemicals to date. The really outstanding fungicide that will be as effective as other forms of biotoxicants remains to be developed.

There is no reason why superior materials cannot be developed because knowledge is rapidly becoming available to guide the research effort away from blind probing and wild conjectures which had to prevail in the early days of the modern era of organic fungicides. The foundations have been laid for the young science of fungous toxicology which should grow and flourish in the next fifty years more than it has in the past.

The principles of how to design molecules for specific attributes are rapidly being established. With all the versatile possibilities of organic chemistry to draw upon, the plant pathologist can look forward with confidence to resolving the problems still before him. The careful research in fungous toxicology has pointed the way, and, if pursued diligently in the future, will be the means of meeting the practical problems as they arise.

In the foregoing story of chemicals at work in agriculture for the prevention of losses caused by destructive plant diseases, man's progress from fumbling in the dark to precise methods for selecting, developing, and adapting safe and specific remedies for the control of particular diseases has been summarized. It is hoped that future generations of plant pathologists will be permitted to continue waging the Battle of the Blights for the benefit of modern societies and the peoples in less fortunate regions of the hemisphere.

BIBLIOGRAPHY

ANON. 1934. Sulfur. An essential to industry and agriculture. A treatise on the properties and applications of sulfur. Texas Gulf Sulfur Co. Bull. New York.

BROOKS, F.T., and M.A. BAILEY. 1919. Silver-leaf disease III. J. Agric. Sci. 9: 189-215.

CARSON, R. 1962. Silent Spring. The Riverside Press, Boston.

CUNNINGHAM, H.S., and E.G. SHARVELLE. 1940. Organic seed protectants for Lima Beans. Phytopathology 30: 4.

DEBARY, A. 1853. Untersuchungen uber der Brandpilze und die durch sie verursachten Krankheiten der Pflanzen mit Rucksicht auf das Getreide und andere Nutzpflanzen. Berlin.

FILTER, P.L. 1977. Regulation. How much is too much? Golf Superintendent. Nov-Dec. p. 31-32.

FORSYTH, W. 1802. A Treatise on the Culture and Management of Tree Fruits. Nichols Publishing Co., London.

GOLDSWORTHY, M.C., and E.L. GREEN. 1933. Some promising fungicides. Phytopathology 23: 561-562.

HAMILTON, J.M., and D.H. PALMITER. 1943. Fermate—a promising new fungicide. Farm Res. (New York) 9.14.

HARRINGTON, G.E. 1941. Thiuram sulfide for turf diseases. Science 93: 311.

HESTER, W.F. 1943. Fungicidal Composition. U.S. Patent 2,317,765.

HEUBERGER, J.W., and D.O. WOLFENBARGER. 1944. Zinc dimethyldithiocarbamate and the control of early blight and anthracnose on tomatoes, and of leafhopper and early blight on potatoes. Phytopathology 34: 1003.

HORSFALL, J.G. 1932. Red oxide of copper as a dust fungicide for combating damping-off by seed treatment. N.Y. Agric. Exp. Stn. Bull. *615*.

HORSFALL, J.G. 1945. Fungicides and Their Action. Chronica Botanica, Waltham, Mass.

HORSFALL, J.G. 1956. Principles of Fungicidal Action. Chronica Botanica, Waltham, Mass.

HORSFALL, J.G., R.W. MARSH, and H. MARTIN. 1937. Studies upon the copper fungicides IV. The fungicidal value of the copper oxides. Ann. Appl. Biol. 24: 867-882.

HOWARD, F.L. 1941. Antidoting toxin of *Phytophthora cactorum* as a means of plant disease control. Science 94: 345.

HUNT, G.M., and G.A. GARRATT. 1938. Wood Preservation. McGraw-Hill Book Co., New York.

LARGE, E.C. 1940. The Advance of the Fungi. Dover Publications, New York.

LODEMAN, E.G. 1896. The Spraying of Plants. Macmillan Book Co., New York.

MCCALLAN, S.E.A. 1930. Studies of fungicides II. The solvent action of spore excretions and other agencies on protective copper fungicides. Cornell Agric. Exp. Stn. Memoir 128: 8-24. Ithaca, N.Y.

MCCALLAN, S.E.A. 1967. History of Fungicides, Vol. I. Torgeson Fungicides. Academic Press, New York.

MILLARDET, P.A. 1885. Traitement du mildiou et du rot. J. Agric. Prat. (Paris) 49: 801-805. (English translation by F.J. Schneiderhan. 1933. Phytopathol. Classic 3: 7-11.)

MILLARDET, P.A. 1885. Sur l'histoire du traitement du mildiou par le sulfate de cuivre. J. Agric. Prat. (Paris) 49: 801-805. (English translation by F.J. Schneiderhan. 1933. Phytopathol. Classic 3: 18-25.)

MILLARDET, P.A., and U. GAYON. 1887. Recherches nouvelles sur l'action que les preparations cuivreuses execent sur le Perenospora de la vugne. J. Agric. Prat. (Paris) 51: 123-129, 156-161.

MULLER, A. 1926. Die innere Therapie de Pflanzen. Monograph. Angew. Entomol. 8: 1-206.

NIKITIN, A.A. 1937. Zeolitic Copper Compounds as Fungicides. Ph.D. Dissertation. Columbia University, New York.

PIERCE, N.B. 1900. Peach leaf curl, its nature and treatment. USDA Div. Veg. Phys. Pathol. Bull. *20.*

PRÉVOST, P. 1807. Memoire sur la cause immediate de la carie ou Charbon des Bles et de plusieurs autres Maladies des Plantes, et sur les Preservatifs de la Carie. Bernard, Paris. (English translation by G.W. Keitt. 1939. Phytopathol. Classic 6: 1-95.)

REES, E. (Translator). 1963. The Iliad by Homer. Modern Library, New York.

ROBERTSON, J. 1821. On the mildew and some other diseases incident to fruit trees. Trans. Hort. Soc. London 5: 175-185.

SESSIONS, A.C. 1936. Fungicide adjustment. Ind. Eng. Chem. 28: 287-290.

TERHORST, W.P., and E.L. FELIX. 1943. 2, 3-Dichloro-1,4-naphthoquinone: A potent organic fungicide. Ind. Eng. Chem. 35: 1255-1259.

TILLET, M. 1937. Dissertation sur la cause qui corrompt et noircit les grains de ble dans les epis; et sur les moyens de prevenir ces accidens. Translation by H.B. Humphrey. Phytopathol. Classic 5: 1-191.

TORGESON, D.C. 1967. Fungicides. An Advanced Treatise. Academic Press, New York.

WEIGHTON, D. 1814. On destroying insects and removing mildew and canker in fruit trees. Memoirs Caledonia Soc. 1: 131-142.

WELLMAN, R.H., and S.E.A. MCCALLAN. 1946. Glyoxalidine derivatives as foliage fungicides. Contrib. Boyce Thompson Inst. 14: 151-160.

WOOLMAN, H.M., and H.B. HUMPHREY. 1924. A summary of literature on bunt or stinking smut of wheat. U.S. Dep. Agric. Bull. *210*.

5

Identity and Nomenclature of Modern Fungicides

With the introduction of organic chemicals for plant protection (circa 1940), a host of different chemicals have paraded across the stage of plant protection under a variety of names and designations. Some of these have survived to assume an important role in the prevention of losses caused by plant diseases, insect pests, weeds, and other unwanted saboteurs of growing plants. Many others have fallen by the wayside to be replaced by more effective compounds, and still others have fallen victim to economic failure or have been banned by governmental restrictions on pesticidal chemicals.

Lack of uniformity in the naming of agricultural pesticides in the U.S.A. and other countries in the early days of organic plant protectants resulted in considerable confusion in the identification of modern pesticides. The use of a multiplicity of trade names, and the assignment of various common names in different countries was confusing even to those directly involved in the use of pesticidal chemicals.

In the early days of the introduction of organic chemicals for plant disease control, plant pathologists recognized the need for some standardized system of nomenclature for commercial fungicides. To most people not directly involved in the development and application of modern pesticides, the materials used for plant disease and insect control are highly complex chemicals with unpronounceable names and a confusing variety of trade names. In 1947 the American Phytopathological Society established a sub-committee on "Standardized Fungicide Nomenclature" to work in cooperation with the U.S. Department of Agriculture, Interdepartmental Committee on Pest Control, to develop a universal system for naming chemicals widely used for plant disease control. At that time, it was proposed that all fungicides with the same active chemical ingredient be designated by "generic," or "common," or "coined names." In 1947 the American Phytopathological Society committee on Fungicide

Nomenclature proposed four "common names" for fungicides in wide usage at that time. These were ferbam, thiram, ziram, and zineb (common names are not capitalized). Thus, "ferbam" included all commercially available fungicide formulations containing ferric dimethyldithiocarbamate ($C_9H_{18}FeN_3S_6$), also designated by commercial trade names such as Fermate, Coromate, Karbam Black, and Ferberk (trade names are capitalized).

Similarly, all fungicidal pesticides containing zinc ethylenebisdithiocarbamate ($C_4H_6N_2S_4Zn$), and distributed under trade names such as Dithane Z-78, Zimate, and Blitox were designated by the common name "Zerlate." "Thiram" is the common name for all products containing Tetramethylthiuram disulfide, and "ziram" is the common name for products containing zinc dimethyldithiocarbamate.

In 1955, five additional common names were adopted for fungicides in common usage at that time. They were:

dichlone—2,3-Dichloro-1,4-naphthoquinone, distributed under the trade name Phygon

chloranil—tetrachloro-p-benzoquinone, distributed under the trade name Spergon

captan—N-(Trichloromethylthio)-4-cyclohexene-1,2-dicarboximide, distributed under trade names such as Captan, Orthocide, and Merpan

glyodin—2-heptadecylglyoxalidine acetate, distributed under trade names such as Glyodex, Crag Fruit Fungicide 341, and Glyoxalidine

maneb—manganous ethylenebisdithiocarbamate, with trade names such as Manzate, Tersan-LSR, Polyram-M, and Lonocol-M

In 1956, the functions of the American Phytopathological Society's Fungicide Nomenclature Committee were assigned to the American Standards Association Sectional Committee K62, responsible for approval of common names for pest control chemicals.

Presently, assignment of common names to pesticides in the U.S.A. is regulated by:

The American National Standards Committee K.62
The American National Standards Institute (ANSI)
1430 Broadway
New York, N.Y. 10018

Organizations represented in the K.62 Committee include, among others, the Weed Society of America (WSA), the American Phytopathological Society (APS), the Entomological Society of America (ESA), and the Environmental Protection Agency (EPA).

Outside the U.S.A. common names of pesticidal chemicals are approved by the International Organization of Standards (ISO), in which the U.S.A. is represented by ANSI. The identification of agricultural pesticides is based internationally on the following scheme:

PESTICIDE (Economic Poison)
Trivial Name
Common Name
Chemical Name
Trade Name(s)

According to the International Organization of Standards (ISO), the terms used in the above scheme are defined as follows:

A pesticide or economic poison is defined under the Federal Insecticide, Fungicide, and Rodenticide Act as any substance or mixture of substances intended for preventing, destroying, repelling, or mitigating any insects, rodents, nematodes, fungi, or weeds, or any other forms of life declared to be pests, and any substance or mixture of substances intended for use as a plant regulator, defoliant, or dessicant. As so defined, economic poisons now are generally known as PESTICIDES.

The trivial name is defined as a name in general or common-place usage; e.g., nicotene sulfate, sulfur, carbamate, arsenical, copper sulfate, etc.

Common name is used if no trivial names are commonplace. Other names may be "coined" for the active ingredients of pest control chemicals, including growth regulators, dessicants, and defoliants, the chemical names of which are so long and complicated that they are not easily remembered or are not adaptable to literature or ordinary usage. In the U.S.A. the American National Standard Institute (ANSI) provides a procedure whereby manufacturers, formulators, or interested organizations (public or private) may obtain recognition and acceptance of a proposed common name. To obtain approval a sponsor submits a statement together with an application fee to:

The Secretary, American National Standards Committee K.62
ANSI
1430 Broadway
New York, N.Y. 10018

Outside the U.S.A. common names of pesticides are approved by the International Organization for Standardization (ISO).

Chemical name is the identity of the active ingredient of a pesticide expressed in acceptable standard chemical terms.

Trade name or (trademark), also known as the "brand name" is defined by the Association of American Pest Control Officials, Inc. as "the name, number, trade mark, or designation applied to an economic poison of any particular description by the manufacturer, distributor, importer or vendor thereof. Each economic poison differing in the ingredient statement, analysis, manufacturer, or distributor, name, number, or trademark shall be considered as a distinct or separate brand."

On the basis of the foregoing definitions and regulations, the identity and nomenclature of some fungicides commonly used in the U.S.A. and elsewhere are summarized in Table 5.1. The list of products does not include all fungicidal products, nor are the products listed necessarily approved for use on food crops. The latter is discussed in detail in Chapter 11.

BIBLIOGRAPHY

ANON. 1978. Farm Chemicals Handbook. Meister Publishing Co., Willoughby, Ohio.

AUSTRALIAN DEPARTMENT OF HEALTH. 1974. Pesticides. Third edition. Australian Government Printing Service, Canberra.

THOMSON, W.T. 1978. Agricultural Chemicals Book IV. Fungicides. Thomson Publications, Fresno, Calif.

TABLE 5.1. IDENTITY OF MODERN FUNGICIDES—COMMON NAMES, TRADE NAMES, CHEMICAL NAMES

Trade names are listed in capital letters
Common names are listed in lower case

Compound	Common or Trivial Name	Other Trade Names	Chemical Name
AATECK		*See* thiram	
AATERRA		*See* etridiazol	
ACQUINITE		*See* chloropicrin	
ACTI-AID		*See* cycloheximide	
ACTI-DIONE		*See* cycloheximide	
ACTI-DIONE PM		*See* cycloheximide	
ACTI-DIONE RZ		*See* cycloheximide	
ACTI-DIONE TGF		*See* cycloheximide	
ACTI-DIONE-THIRAM		*See* cycloheximide	
AFUGAN		*See* pyrazophos	
AGRIMYCIN		*See* streptomycin	
AGRIMYCIN 17		*See* streptomycin	
AGRI-STREP		*See* streptomycin	
AGROSAN		*See* PMA	
AGROSOL-S		30% maneb + 30% captan	
AGROX 2 WAY		Combination of captan and diazinon	
AGROX 3 WAY		Combination of captan, Diazinon and Lindane	
ALLISAN		*See* dicloran	
AMOBAM		CHEM-O-BAM	Diammonium ethylene bis-dithiocarbamate Used in a tankmix with zinc sulfate to form zineb
	anilazine	DYRENE, DIREZ, KEMATE, TRIASYN	4,6-Dichloro-N-(2-chloro-phenyl)-1,3,5 triazin-2-amine
ANTI-CARIE		*See* hexachlorobenzene	
ANTRACOL		*See* propineb	
ARASAN		*See* thiram	
ARATHANE		*See* dinocap	
ASOZIN			Methylarsenic sulfide
ASPOR		*See* zineb	
AVICOL		*See* PCNB	
AZIDE		KAZOE, SMITE	Sodium or potassium azide

TABLE 5.1 *(Continued)*

Trade names are listed in capital letters
Common names are listed in lower case

Compound	Common or Trivial Name	Other Trade Names	Chemical Name
BACTICIN		Mixture of 2,4-xylenol and meta-cresol	
BANROT		Mixture of 15% terrazole and 25% TOPSIN M	
BAS 3170		*See* benodanil	
BAS 3191		*See* furcarbanil	
BAS 3460		*See* carbendazim	
BASFUNGIN		*See* methylmetiram	
BASICOP		*See* copper oxychloride	
	basic copper carbonate	CUPRIC CARBONATE, COBREDON, KOP-KARB, MALACHITE	Basic copper carbonate
	basic copper chloride	*See* copper oxychloride	
	basic copper sulfate	BASICOP, COPPER 5 DUST, COPPER HYDROXYSULFATE, COPPER OXYSULFATE, TRI-BASIC, TBCS-53	Basic copper sulfate
	basic cupric zinc sulfate	BSZ, COP-O-ZINC, ZINC COPOSIL	Basic cupric zinc sulfate complex
BATASAN		*See* fentinacetate	
BAVISTIN		*See* carbendazim	
BAY 33172		*See* fuberidazole	
BAY 47531		*See* dichlofluanid	
BAY 78418		*See* edifenphos	
BAY 79770		*See* chloraniformethane	
BCM		*See* carbendazim	
BENLATE		*See* benomyl	
	benodanil	BAS 3170 F	2-iodobenzoic acid anilide
	benomyl	BENLATE, GREX, ULTRA SOFRIL, TERSAN 1991	methyl 1-(butylcarbamoyl)-2-benzimidazolecarbamate
	benquinox	CEREDON, CERELINE, COBH, TILLANTOX, TSERENOX	1,4-Benzoquinone N-benzoylhydrazone oxime
	binapacryl	MOROCIDE	
BIOQUIN 1			copper 8 quinolinate

	biphenyl	DIPHENYL, LEMONENE, PHENADOR-X, PHENYL-BENZENE	diphenyl
BLA-S		*See* blasticidin	
	blasticidin	BLASTICIDIN-S, BLA-S	benzylaminobenzene sulphonate
BLITANE		Mixture of maneb and copper oxychloride	
BLITHANE		Mixture of zineb and copper oxychloride	
BLITOX		*See* copper oxychloride	
BLUE COPPERAS			Copper sulfate pentahydrate
BLUESTONE			Copper sulfate pentahydrate
BLUE VITRIOL			Copper sulfate pentahydrate
BORDEAUX		Mixture of copper sulfate and calcium hydroxide forming basic copper sulfates	
BORDO MIXTURE		*See* BORDEAUX	
BOTEC		Mixture of BOTRAN and captan	
BOTRAN		*See* dicloran	
BOTRILEX		*See* PCNB	
BRASSICOL		*See* PCNB	
BRAVO-W 75		*See* chlorothalonil	
BRESTAN		*See* fentinacetate	
BRIMSTONE		*See* sulfur	
BROMOFUME		*See* ethylene dibromide	
BROM-O-GAS		*See* methyl bromide	
BROMOSAN		Mixture of CERCOBIN and thiram	
BRONOPOL			2-bromo-2-nitropropan-1,3-diol
BROWN COPPER OXIDE		*See* cuprous oxide	
BSZ			Basic cupric zinc sulfate complex
BUNT-NO-MORE		*See* hexachlorobenzene	
BUNTOSAN		Mixture of 75% mancozeb and 0.5% BHC	
BUNTOSAN-G		Mixture of 29% hexachlorobenzene and 0.5% BHC	
	bupirimate	PP 588, NIMROD	5-Butyl-2-ethylamino-6-methylpyrimidin-4yl-dimethyl-sulfamate
BUSAN-30			2-(Thiocyanomethylthio) benzothiazole
BUSAN 72		BUSAN-30, TCMTB	2-(Thiocyanomethyl) benzothiazole

TABLE 5.1 ***(Continued)***

Trade names are listed in capital letters
Common names are listed in lower case

Compound	Common or Trivial Name	Other Trade Names	Chemical Name
BUTAFUME		FRUCOTE, DECCOTANE, TUTANE	2-Aminobutane
CADDY		VI-CAD	Cadmium chloride
CADMINATE			Cadmium succinate
CAD-TRETE		Mixture of 75% thiram and 8.3% cadmium chloride	
CALCIUM HYDROXIDE		*See* hydrated lime	
CALIXIN		*See* tridemorph	
CAPRANE		*See* dinocap	
CAPRYL		*See* dinocap	
	captafol	DIFOLATAN, DIFOSAN, SULFONIMIDE, SANSPOR	Cis-N-(1,1,2,2-tetrachloro-ethylthio)-4-cylclohexene-1,2-dicarboximide
	captan (captane)	CAPTAN, CAPTANE, MERPAN, ORTHOCIDE, VONDCAPTAN	N-(Trichloromethylthio)-4-cyclohexene-1,2-dicarbox-imide
CAPTHION		Mixture of captan, sulfur and malathion	
CARBAMATE		*See* ferbam	
CARBATENE		*See* metiram	
	carbendazim	BCM, BAS 3460, BAVISTIN, DEROSAL	2-(methoxycarbonylamino)-benzimidazole
	carboxin	DCMO, VITAVAX	5,6-Dihydro-2-methyl-1,4-oxathiin-3-carboxanilide
CARPENE		*See* dodine	
CECA		UDONKOR	N-(beta-Cyanoethyl) mono-chloroacetamide
CELA W.524		*See* triforine	
CELDION		*See* fentiazon	
CELLU-QUIN		BIOQUIN 1, CUNILATE, COPPER-8, COPPER OXI-NATE, MILMER, OXIME COPPER, QUINONDO	Copper-8-hyroxyquinolate
CERCOBIN		*See* thiophanate	

CERCOBIN-M		*See* thiophanate methyl	
CERCOBIN METHYL		*See* thiophanate methyl	
CEREDON		*See* benquinox	
CERELINE		*See* benquinox	
CERESAN UNIVERSAL		DYNACIDE, PMA, PMAS, PMACETATE, PHIX, PMAC, LIQUIPHENE, PURATURF NO:10, PURASAN -SC-10	Phenyl Mercuric acetate
CHEM-BAM		*See* nabam	
CHEM-O-BAM		*See* AMOBAM	
CHEMFORM		*See* streptomycin	
CHINOMETHIONAT		*See* oxythioquinox	
CHINOMETHIONATE		*See* oxythioquinox	
	chloranil	SPERGON	2,3,5,6-Tetrachloro-1,4-benzoquinone
	chloroneb	DEMOSAN, TERSAN-SP	1,4-Dichloro-2,5-dimethoxy benzene
CHIPCO SPOT KLEEN		*See* thiophanate	
CHLORO-O-PIC		*See* chloropicrin	
CHLOROPICRIN		ACQUINITE, CHLOR-O-PIC, DOLOCHLOR, LARVACIDE, PICFUME, PIC-CLOR, TRICHLOR	Trichloronitromethane (nitrochloroform)
	chloroquinox	LUCEL	5,6,7,8-Tetrachloroquinoxa-line
	chlorothalonil	BRAVO-W 75, DACONIL 2787, EXOTHERM, TERMIL	Tetrachloroisophthalonitrile
CILUAN		*See* pyridinitril	
CITCOP-4E		COPOLOID, COPPER LINO-LEATE	Copper salts of fatty and rosin acids. 20-25% copper abietate, 8-12% copper linoleate, and 10-15% copper oleate
CNA		*See* dicloran	
CNC	Copper naphthenate		Cupric cyclopentane carboxy-late
COBH		*See* benquinox	

TABLE 5.1 *(Continued)*

Trade names are listed in capital letters
Common names are listed in lower case

Compound	Common or Trivial Name	Other Trade Names	Chemical Name
COBOX		*See* copper oxychloride	
COBOX BLUE		*See* copper oxychloride	
COBREDON			Basic copper carbonate
C.O.C.S.		COPRO 53, COPPER OXY-CHLORIDE SULFATE	Mixture of basic copper chloride and basic copper sulfate
COLLOIDOX		*See* copper oxychloride	
CONSUL			Elemental sulfur
COP-O-CIDE		*See* copper resinate	
COPOLOID		*See* CITCOP	
COPOSIL		*See* Basic cupric zinc sulfate complex	
COPOX			Cuprous oxide
COP-O-ZINC			Basic cupric Zinc sulfate complex
COPPER 5 DUST		*See* Basic copper sulfate	
	Copper-8-Quinolinate	BIOQUIN, COPPER OXI-NATE, COPPER -8, CELLU-QUIN, CUNILATE, MILMER, OXIME COPPER	Copper-8-quinolinolate (Copper-8 hydroxyquino-late
	Copper ammonium carbonate	*See* Copper ammonium complex	
COPPER-COUNT-N		*See* Copper ammonium complex	
COPPER-CURE			Cupric cyclopentanecarboxy-late
COPPERFINE-ZINC			Copper sulfate pentahydrate
COPPER-FIXED	Includes the basic sulfate oxychlorides and oxides of copper		
	Copper hydroxide	CUPRAVIT BLUE, KOCIDE 101, KOCIDE 220 and KOCIDE 404	Copper hydroxide
COPPER LIME DUST			Monohydrated copper sulfate and Lime in 20-80 propor-tion
COPPER LINOLEATE		*See* CITCOP-4E	

COPPERNATE			Cupric cyclopentanecarboxylate
COPPER OC FUNGICIDE		*See* copper oxychloride	
COPPER OLEATE			Organic copper compound
COPPER OXIDE			Cuprous oxide
	copper oxychloride	BLITOX, COBOX, COBOX BLUE, COLLOIDOX, COPPER OC FUNGICIDE, COPRANTOL, COPPERSAN, COXYSAN, CUPROKYLT, CUPROXOL, CUPROVINOL, CUPROX, CUPRAMAR, CUPRAVIT, MICROCOP, KAURITIL, VIRICUIVRE	Copper oxychloride
COPPER OXYCHLORIDE SULFATE		C-O-C-S, COPRO-53	A mixture of basic copper chloride and basic copper sulfate
COPPER OXINATE			Copper-8-quinolinolate
COPPER OXYSULFATE		*See* Basic copper sulfate	
	copper resinate	COP-O-CIDE	An emulsion of copper salts of fatty and rosin acids
COPPERSAN			Copper oxychloride
COPPER-SARDEZ			Cuprous oxide
Copper Sulfate monohydrate			Copper sulfate pentahydrate
	copper lime mixture	NEUTROCOP, NU-COP, SPRAY-COP	Copper lime mixtures
COPPER UVERSOL	copper naphthenate		Cupric cyclopentanecarboxylate
COPPER WHITE			Mixture of copper and zinc
COPRANTOL		*See* copper oxychloride	
COP-R-NAP	copper naphthenate		Cupric cyclopentanecarboxylate
COPRO 53		*See* C.O.C.S.	
COROZOATE		*See* ziram	
COSAN			Elemental sulfur
COSANIL		Mixture of 25% zineb, 20% sulfur and 5% copper oxychloride	
COXYSAN		*See* copper oxychloride	
CRAG TURF FUNGICIDE 531			Cadmium calcium, copper, zinc chromate complex

TABLE 5.1 *(Continued)*

Trade names are listed in capital letters
Common names are listed in lower case

Compound	Common or Trivial Name	Other Trade Names	Chemical Name
CROTOTHANE		*See* dinocap	
CUFRAM-Z	Dithiocarbamate complex containing copper, manganese, iron and zinc		
CUFRANEB		Same as CUFRAM-Z	
CUMAN		*See* ziram	
CUNILATE			Copper-8 quinolinolate
COPRANTOL		*See* Copper oxychloride	
CUPRAVIT BLUE		*See* Copper hydroxide	
CUPRAVIT GREEN		*See* Copper hydroxide	
CUPRIC CARBONATE	basic copper carbonate	CUPRIC CARBONATE, MALACHITE, COBREDON, KOP KARB	Basic copper carbonate
CUPRINOL BROWN	copper naphthenate		Cupric cyclopentanecarboxylate
CUPROCIDE	cuprous oxide		Cuprous oxide
CUPROKYLT		*See* Copper oxychloride	
CUPROL		*See* Copper oxychloride	
CUPROSAN BLUE		*See* Copper oxychloride	
CUPROVINOL		*See* Copper oxychloride	
CUPROX		*See* Copper oxychloride	
CUPROXOL		*See* Copper oxychloride	
CURAMIL		*See* pyrazophos	
CURITAN		*See* dodine	
	cycloheximide	ACTI-AID, ACTI-DIONE, ACTI-DIONE PM, ACTI-DIONE RZ, ACTI-DIONE TGF, ACTI-DIONE THIRAM, ACTI-SPRAY, NARAMYCIN, KAKEN, HIZAROCIN	3 (2-(3,5-Dimethyl-2-oxocyclohexyl)-2-hydroxyethyl)-glutarimide
CYCLOMORPH		*See* dodemorph	
	cypendazole	DAM 18654, FOLCIDIN	1-(5-cyanopentylcarbamoyl-2-(methyloxycarbonylamino)-benzimidazole
CYPREX		*See* dodine	

DACONIL		*See* chlorothalonil	
DAPA		*See* fenaminosulf	
DCMO		*See* carboxin	
DCMOD		*See* oxycarboxin	
DCNA		*See* dicloran	
DDPP		*See* pyridinitiril	
DECCOTANE	2-aminobutane	BUTAFUME, FRUCOTE, TUTANE	2-aminobutane
DELAN		*See* dithianon	
DELAN-COL		*See* dithianon	
DEMOSAN		*See* Chloroneb	
DEMOSAN-T		40% DEMOSAN & 22.5% thiram	
DEROSAL		*See* carbendazim	
DEXON		*See* fenaminosulf	
DIBAM		*See* nabam	Sodium dimethyl dithiocarbamate
	dichlofluanid	BAY 47531, ELVARON, EUPARENE, EUPAREN	N-Dichlorofluoromethylthio-N-N-dimethyl-N-phenylsulfamide
	dichlone	PHYGON	2,3-Dichloro-1,4-naphthoquinone
	dichlorophen (antiphen, dichlorophene)	PREVENTOL	Di-(5-chloro-2-hydroxyphenyl) methane
	dicloran	ALLISAN, BOTRAN, CNA, DCNA, DITRANIL, RESISAN	2,6-Dichloro-4-nitroaniline
	dichloropropene	TELONE II	1,3-Dichloropropene
DIFOLATAN		*See* captafol	
DIFOSAN		*See* captafol	
DIKAR	A coordination product of zinc ion and manganese ethylenebisdithiocarbamate		
	dimethirimol	MILCURB	5-n-Butyl-2-dimethylamino-4-hydroxy-6-methylpyrimidine
	dinocap	CAPRANE, CAPRYL, CROTOTHANE, DNOCP, ISCOTHAN, KARATHANE, MILDEX	2,4-Dinitro-6-octyl-phenylcrotonate + 2,6-dinitro-4-octyl phenyl crotonate + nitro-octyl phenols
DIPHENYL		*See* biphenyl	
DIREZ		*See* anilazine	

TABLE 5.1 *(Continued)*

Trade names are listed in capital letters
Common names are listed in lower case

Compound	Common or Trivial Name	Other Trade Names	Chemical Name
	ditalimfos	DOWCO 199, LAPTRAN, PLONDREL	0, 0-diethyl phthalimido-phosphonothioate
DITHANE D-14		*See* nabam	
DITHANE A-40		*See* nabam	
DITHANE M-22		*See* maneb	
DITHANE M-22 SPECIAL		A mixture of maneb and zinc sulfate	
DITHANE M-45		*See* mancozeb	
DITHANE S-31		A mixture of 53% maneb and 19% nickel sulfate	
	dithianon	DELAN, DELAN-COL, THYNON	5,10-Dihydro-5,10-dioxo-naphtho-(2,3,9)-p-dithiin-2,3-dicarbonitrile
DITRANIL		*See* dicloran	
DNC		*See* DNOC	
DNOC		DNC, ELGETOL 30, NITRA-DOR, SELINON, SINOX, TRI-FOCIDE	4,6-Dinitro-o-cresol
DNOCP		*See* dinocap	
	dodemorph (dodemorfe)	CYCLOMORPH,MELTATOX	N-Cyclododecyl-2,6-dimethylmorpholinium acetate
	dodine	CYPREX, CARPENE, CURI-TAN, DOQUADINE, MEL-PREX, SYLLIT, VONDODINE	n-Dodecylguanadine acetate
DOLOCHLOR		*See* chloropicrin	
DOWCO 199		*See* ditalimfos	
DOWCO 269		*See* pyroxychlor	
DOWFUME 75		70% ethylene dichloride + 30% carbon tetrachloride	
DOWFUME EB-5		29.2% ethylene dichloride + 63.6% carbon tetrachloride	
DOWFUME MC-2		98% methyl bromide + 2% chloropicrin	
DOWFUME MC-33		67% methyl bromide + 33% chloropicrin	
DOWFUME N		50% 1,3-Dichloropropene + other chloropropenes	

DOWFUME W-85		83% ethylene dibromide	
	drazoxolon	GANOCIDE, MIL-COL, PP 781, SOPRACOL	4-(2-chlorophenylhydrazono)-3-methyl-5-isoxazolone
DOWICIDE-A			Sodium-o-phenyl-phenate
DOWICIDE 1			ortho-phenyl phenol
DOWICIDE 2			2,4,5-trichlorophenol
DOWICIDE 7			Pentachlorophenol
DOWICIDE G			Pentachlorophenol
DRB			2,4-Dinitrophenyl thiocyanate
DSE		*See* nabam	
DU-TER		*See* fentin hydroxide	
DYNACIDE			Phenyl Mercuric Acetate
DYRENE		*See* anilazine	
EARTHCIDE		*See* PCNB	
EAU GRISON		*See* LIME-SULFUR	
EDDP		*See* edifenphos	
	edifenphos	BAY 78418, EDDP, HINOSAN, LUTROL	O-Ethyl-S,S,-diphenyl-dithiophosphate
EL-291		*See* tricyclazole	
ELGETOL 30		*See* DNOC	
ELVARON		*See* dichlofluanid	
ENOVIT		*See* thiophanate	
ENOVIT-SUPER		*See* thiophanate methyl	
	etridiazol	KOBAN, AATERRA, TERRAZOLE, TRUBAN, ETCMB, ETHAZOL	5-Ethoxy-3-trichloromethyl-1,2,4-thiadiazole
	ethirimol	MILCURB, MILGO E, MILSTEM, NEW MILSTEM, SUPER MILGO	5-Butyl-2-ethylamino-4-hydroxy-6-methylpyrimidin
EUPAREN		*See* dichlofluanid	
EUPARENE		*See* dichlofluanid	
EXOTHERM		*See* chlorothalonil	
F.461		*See* oxycarboxin	
	fenaminosulf	DEXON	
FENNITE-A		Mixture of 6.7% tri-phenyltin acetate and maneb	
	fentiazon	CELDION	3-Benzylideneamino-4-phenylthiazole-2-thione
FENOLOVO		*See* fentin hydroxide	
FENTIN		*See* fentinacetate	

TABLE 5.1 *(Continued)*

Trade names are listed in capital letters
Common names are listed in lower case

Compound	Common or Trivial Name	Other Trade Names	Chemical Name
	fentinacetate	BRESTAN, BATASAN, FENTIN, PHENTINOACETATE, TPTA, TUBOTIN	triphenyltin acetate
	fentin hydroxide	DU-TER, FENOLOVO, TPTH, TPTOH	triphenyltin hydroxide
FERBAME		*See* ferbam	
	ferbam	CARBAMATE, FERBERK, FERMOCIDE, FERBAME, FERMATE, HEXAFERB, HOKMATE, NIACIDE, TRIFUNGOL, VANCIDE FE 95	Ferric dimethyldithiocarbamate
FERBERK		*See* ferbam	
FERMATE		*See* ferbam	
FERMIDE		*See* thiram	
FERMOCIDE		*See* ferbam	
FERNACOL		*See* thiram	
FERNASAN		*See* thiram	
FOLCIDIN			Methyl-1-(5-cyanopentylcarbamoyl) benzimidazole-2-ylcarbamate
FOLPAN		*See* folpet	
	folpet	FOLPAN, PHALTAN, THIOPHAL	N-(Trichloromethylthio) phthalimide
FORE		*See* mancozeb	
FORSTAN		*See* oxythioquinox	
FRUCOTE		2AB, BUTAFUME, DECCOTANE, 2-AMINOBUTANE	2-aminobutane
	fuberidazole	BAY 33172, FURIDAZOLE, NEO-VORONIT, VORONIT SPECIAL	2-(2″-furyl)-benzimadazole
FUCLASIN ULTRA		*See* ziram	
FUKLASIN		*See* ziram	
FULCIN		*See* griseofulvin	

FULVICIN		*See* griseofulvin	
FUNGINEX		*See* triforine	
FUNGO		*See* thiophanate methyl	
	furcarbanil	BAS 3191	2,5-Dimethylfuran-3-carboxanilide
FUNGICLOR		*See* PCNB	
FURIDAZOL		*See* fuberidazole	
GALLOTOX		*See* PMA	
GANOCIDE		*See* drazoxolon	
GEROX		*See* streptomycin	
	glyodin	CRAG FRUIT FUNGICIDE 341	2-Heptadecyl-2-imidazoline acetate
GRANOL-N-M		Mixture of 50% maneb and 18.75 BHC	
GRANOX LIQUID		Mixture of 25% maneb and 5% hexachlorobenzene	
GRANOX-N-M		Mixture of 50% maneb and 10% hexachlorobenzene	
GRANOX-P.F.M		Mixture of 30% maneb and 30% captan	
GRESFEED		*See* griseofulvin	
GREX		*See* benomyl	
GRIFULVIN		*See* griseofulvin	
	griseofulvin	FULCIN, FULVICIN, GRIFULVIN, GRESFEED, GRISETIN, MURFULVIN	7-Chloro-4,6-dimethoxycoumaran-3-one-2-spiro-1-[2-methoxy-6-methylcyclohex-2-en-4-one]
GRISETIN		*See* griseofulvin	
	guazatine	GUANOCTINE, MC-25, PANOCTINE	9-aza-1,17-diguanidinoheptadecane salt (either sulfate or acetate)
HCB			Hexachlorobenzene
HEXA-C.B		*See* hexachlorobenzene	
	hexachlorobenzene	ANTI-CARIE, BUNT-NO-MORE, HCB, HEXA-C.B., NO BUNT 40, SANOCIDE, SMUT-GO HEXIDE	Hexachlorobenzene
HEXACHLOROPHENE		ISCOBAC, NABAC	Sodium salt of 2, 2-methylenebis (3,4,6-trichlorophenol)

TABLE 5.1 *(Continued)*

Trade names are listed in capital letters
Common names are listed in lower case

Compound	Common or Trivial Name	Other Trade Names	Chemical Name
HEXAFERB		*See* ferbam	
HEXASUL	elemental sulfur		Elemental sulfur
HEXATHANE		*See* zineb	
HEXATHIR		*See* thiram	
HEXAZIR		*See* ziram	
HEXIDE		*See* hexachlorophene	
HINOSAN		*See* edifenphos	
HIZAROCIN		*See* cycloheximide	
HOE 2873		*See* pyrazophos	
HOKKO-MYCIN		*See* streptomycin	
HOKMATE		*See* ferbam	
HYDRATED LIME	slaked lime		Calcium hydroxide
HYDROXYISOXAZOLE		HYMEXAZOL, TACHIGAREN	3-hydroxy-5-methylisoxa-zole
HYMEXAZOL		*See* HYDROXYISOXAZOLE	
IBP		KITAZIN, KITAZIN-P	S-Benzyl-O-O-diisopro-pyl-phosphorothioate
IMUGAN		*See* chloraniformethane	
INDAR			4-N-Butyl-1,2,4-triazole
ISOCOTHAN		*See* dinocap	
ISOBAC			Sodium salt of 2,2-methyl-enebis (3,4,6-trichloro-phenol)
ISOTOX SEED TREATMENT		Mixture of 12.5% captan and 25.0% Lindane	
KARATHANE		*See* dinocap	
KAURITIL		*See* copper oxychloride	
KEMATE		*See* anilazine	
KITAZIN		KITAZIN-P, IBP	S-Benzyl-O-O-diisopro-pyl-phosphorothioate
KOBAN		*See* etridiazol	

KOBUTOL		*See* PCNB	
KOCIDE 101		COPPER HYDROXIDE, CUPRAVIT BLUE	Copper hydroxide
KOCIDE 220			Copper hydroxide
KOCIDE 404			Copper hydroxide
KOLOFOG	sulfur		Elemental sulfur
kolospray	sulfur		Elemental sulfur
KOP KARB		BASIC COPPER CARBONATE, CUPRIC CARBONATE MALACHITE	Basic copper carbonate
KROMAD	Mixture of 16% thiram, 5% Cadmium Sebacate, 5% Potassium chromate, 1% malachite and 5% Auramine		
KUMULUS	sulfur		Elemental sulfur
KUPRITE		*See* cuprous oxide	
KYPMAN-80		*See* maneb	
KYPZIN		*See* zineb	
LAPTRAN		*See* ditalimfos	
LARVACIDE		*See* chloropicrin	
Lauxtol A		Pentachlorophenol	
LEMONENE		*See* biphenyl	
LEYTOSAN			Phenyl Mercuric Acetate
LIME-SULFUR			Calcium Polysulfide
LIQUIPHENE			Phenyl Mercuric Acetate
LONACOL		*See* zineb	
LONACOL M		*See* maneb	
LORVEK		*See* pyroxychlor	
LUCEL		*See* chlorquinox	
LUTROL		*See* edifenphos	
MAGNETIC 70			Elemental Sulfur
MALACHITE			Basic Copper Carbonate
MANCOFOL		*See* mancozeb	
	maneb (manebe)	DITHANE M-22, DITHANE M-22 SPECIAL, KYPMAN 80, MANEB 80, MANEBA, MANESAN, MANZATE, MANZATE-D, MANZIN, MANOC, MANEBGAN, POLYRAM-M, REMASAN,	Manganese ethylene bisdithiocarbamate

TABLE 5.1 *(Continued)*

Trade names are listed in capital letters
Common names are listed in lower case

Compound	Common or Trivial Name	Other Trade Names	Chemical Name
MANEB (Contd.)		SOPRANEBE, TERSAN-LSR, TRIMANGOL, TRIMANOC, TUBOTHANE	
MANEB-80		*See* maneb	
MANEBA		*See* maneb	
MANEBGAN		*See* maneb	
MANESAN		*See* maneb	
MANOC		*See* maneb	
MANZATE		*See* maneb	
MANZATE-200		*See* mancozeb	
MANZATE-D		*See* maneb	
MANZIN		*See* maneb	
MAS			Methylarsenic sulfide
MC-25		*See* guazatine	
MEB		*See* maneb	
MEBENIL		*See* methylmetiram	
MELPREX		*See* dodine	
MELTATOX		*See* dodemorph	
MEMC	MEMC	AGALLOL, ARETAN 6, BAYTAN, CELMER, EMISAN 6, TAYSSATO	Methoxyethylmercury chloride
MERPAN		*See* captan	
MERTECT		*See* thiabendazole	
	methyl arsenic sulfide	ASOZIN, MAS, RHIZOCTOL, URBASULF	Methylarsenic sulfide
	methylmetiram	BASFUNGIN, MEBENIL	O-methyl benzoic acid anilide
	metiram	CARBATENE, POLYRAM, PETD, POLYRAM COMBI, ZINC METIRAM	Coprecipitation of zinc ammonium ethylene bisdithiocarbamate and polyethylene bisdithiocarbamoyl disulfide

MEZENE		*See* ziram	
MEZINEB		*See* propineb	
MICROCOP		*See* copper oxychloride	
MICROFLOTOX			Elemental sulfur
MILBAM		*See* ziram	
MIL-COL		*See* drazoxolon	
MILCURB		*See* dimethirimol	
MILDEX		*See* dinocap	
MILDOTHANE		*See* thiophanate methyl	
MILFARON		*See* chloraniformethane	
MILGO		*See* ethirimol	
MILMER	Organic copper	COPPER OXINATE, BIO-QUIN, CUNILATE, CELLU-QUIN, QUINONDO	Copper-8-quinolinolate
MILSTEM		*See* ethirimol	
MONZET		URBACID, URBACIDE	Methylarsenic-dimethyl-dithiocarbamate
MORESTAN		*See* oxythioquinox	
MOROCIDE = BINOPACRYL			
MURFULVIN		*See* griseofulvin	
MYCOLZOL		*See* thiabendazole	
NABAC		*See* hexachlorophene	2, 2-methylene bis (3,4,6-tri-chloro-phenol)
	nabam	CHEM-BAM, DITHANE D-14, DSE, PARZATE, SPRING-BAK	Disodium ethylene-1,2-bisdithiocarbamate
NABAME		*See* nabam	
NAPTOX	Copper naphthenate		Cupric cyclopentanecar-boxylate
NEO-SO-SIN-GIN	Neo asozin		Ferric monomethylarson-ate
NEO-VORONIN		*See* fuberidazole	
NEUTROCOP	Copper sulfate		Cupric sulfate monohydrate
NEW MILSTEM		*See* ethirimol	
NIACIDE		*See* ferbam	
NIMROD		*See* bupirimate	
NITRADOR		*See* DNOC	
NO-BUNT 40		*See* hexachlorobenzene	
NOMERSAN		*See* thiram	
NU-COP	Copper sulfate		Copper sulfate monohy-drate

TABLE 5.1 *(Continued)*

Trade names are listed in capital letters
Common names are listed in lower case

Compound	Common or Trivial Name	Other Trade Names	Chemical Name
NURELLE		*See* pyroxychlor	
OLEOCUIVRE	Copper oxide		Cuprous oxide
ORTHOCIDE		*See* captan	
ORTHORIX	Lime-sulfur		Calcium polysulfide
ORTHOXENOL	Phenyl phenol	DOWICIDE 1, TORSITE	o-Phenyl phenol
	Oxycarboxin	DCMOD, F.461, PLANTVAX	
OXIME-COPPER	Copper naphthenate	MILMER, COPPER OXINATE, BIOQUIN, CUNILATE, CELLU-QUIN, QUINONDO	Copper-8-quinolinolate
OXYCOP 8L		SOL KOP 10, TAYTOX, COPPER-COUNT-N	Copper ammonium complex
	Oxythioquinox	FORSTAN, CHINOMETHIONATE, MORESTAN, QUINOMETHIONATE	6-Methyl-1-3-dithiolo (4, 5-b) quinoxalin-2-one
PANOCTINE		*See* guazatine	
PARINOL	parinol	PARNON	a,a-Bis (p-chlorophenyl)-3-pyridine-methanol
PARZATE		*See* nabam	
PCNB	quintozene	AVICOL, BOTRILEX, BRASSICOL, EARTHCIDE, FOLSAM, FUNGICLOR, PENTAGEN, Tilcarex, TRITISAN	Pentachloronitrobenzene
PCP			Pentachlorophenol
PELT-44		*See* thiophanate methyl	
PELT SOL		*See* thiophanate	
PENCHLOROL		Pentachlorophenol	
PENTA	penta		Pentachlorophenol
PENTAGEN		*See* PCNB	
PENTA-KIL			Pentachlorophenol
PENTANOL			Pentachlorophenol
PENTAPHENATE		SODIUM PENTACHLOROPHENATE	Sodium pentachlorophenate

PENTASOL			Pentachlorophenol
PERATOX			Pentachlorophenol
PERCHLOROBENZENE		*See* hexachlorobenzene	
PERECOT	Copper oxide	PERENOX, CUPROCIDE, KUPRITE, OLEOCUIVRE, COPOX	Cuprous oxide
PERENOX			Cuprous oxide
PERMACIDE			Pentachlorophenol
PERMITE			Pentachlorophenol
PETD		*See* metiram	
PHALTAN	folpet	FOLPAN, THIOPHAL	N-(Trichloromethylthio) phthalimide
PHENADOR-X		*See* biphenyl	
PHENAZIN			Phenazin-5-oxide
PHENTINO-ACETATE		*See* fentinacetate	
PHENYLBENZENE		*See* biphenyl	
PHENYL PHENOL			o-Phenyl Phenol
PHIX		*See* PMA	
PHYGON		*See* dichlone	
PHYTOMYCIN		*See* streptomycin	
PIC-CLOR		*See* chloropicrin	
PICFUME		*See* chloropicrin	
PIPRON		*See* piperalin	
PLANTVAX		*See* oxycarboxin	
PLONDREL		*See* ditalimfos	
	PMA	CERESAN UNIVERSAL, AGROSAN, DYNACIDE, GALLOTOX, LIQUIPHENE, MERSOLITE, LEYTOSAN, PHIX, PMAS, PMAC, PURASIN-Sc-10, PURATURF No:10	Phenyl Mercuric Acetate
POLYOXIN			Antibiotic
POLYRAM		*See* metiram	
POLYRAM-M		*See* maneb	
POLYRAM-ULTRA		*See* thiram	
POMARSOL		*See* thiram	
POMASOL-Z		*See* zineb	
PP 588		*See* bupirimate	
PREVENTOL		*See* dichlorophen	

TABLE 5.1 *(Continued)*

Trade names are listed in capital letters
Common names are listed in lower case

Compound	Common or Trivial Name	Other Trade Names	Chemical Name
PREVICUR		*See* prothiocarb	
PRODARAM		*See* ziram	
	propineb	ANTRACOL, PROPINEBE	Zinc propylene bisdithio-carbamate
	prothiocarb	PREVICUR, SN-41703	S-ethyl-N-(3-dimethyla-minopropyl)-thiol car-bamate hydrochloride
PURASAN-SC-10			Phenyl Mercuric Acetate
PURATURF No:10			Phenyl Mercuric Acetate
	pyrazophos	AFUGAN, CURAMIL, HOE 2873	O,O-Diethyl-O-(5-methyl-6-ethoxy-carbonyl-pyra-zole-(1,5-a) pyrimid-2-yl)-thionophosphate
	pyridinitril	CILUAN, DDPP	2,6-Dichloro-3,5-dicyano-4-phenylpyridine
	pyracarbolid	SICAROL, HOE 13764	2-methyl-5,6-dihydro-4-H-pyran-3-carboxylic acid anilide
	pyroxychlor	DOWCO 269, LORVEC, LORVEK, NURELLE	2-chloro-6-methoxy-4-(trichloromethyl) pyridine
QUICKSAN 20			Phenyl Mercuric Acetate
QUINOMETHIONATE		*See* oxythioquinox	
RABCIDE	phthalide compound		4,5,6,7 Tetrachlorophthalide
REMASAN		*See* maneb	
RESISAN		*See* dicloran	
RHIZOCTOL		ASOZIN, MAS, URBASULF	Methylarsenic sulfide
RIMOCIDIN		*See* streptomycin	

ROT-NOT	copper naphthenate		Cupric cyclopentanecarboxylate
RP-26019	hydantoin compound		1-isopropyl carbamoyl-3 (-3,5-dichlorophenyl) hydantoin
SANOCIDE	chlorinated hydrocarbon	ANTI-CARIE, BUNT-NO-MORE, HCB, SMUT-GO	hexachlorobenzene
SANSPOR		*See* captafol	
SAPROL	triforine	CELA W 524, FUNGINEX	N,N-(piperazinediylbis (2,2,2-trichloroethylidene) bis (formamide))
SELINON		*See* DNOC	
SICAROL		*See* pyracarbolid	
SINOX		*See* DNOC	
SMITE			Sodium Azide
SMUT-GO		*See* hexachlorobenzene	
SN-41703		*See* prothiocarb	
SODIUM PENTACHLOROPHENATE		PENTAPHENATE	Sodium pentachlorophenate
SOFRIL	sulfur		Elemental sulfur
SOL KOP 10		OXY COP 8L, COPPER-COUNT-N, TAYTOX	Copper Ammonium Complex
SOPP		DOWICIDE A, TOPANE	Sodium-o-Phenyl Phenate
SOPRACOL		*See* drazoxolon	
SOPRANEBE		*See* maneb	
SPERGON		*See* chloranil	
SPERLOX-S	sulfur		Elemental sulfur
SPERLOX-Z		*See* zineb	
SPERSUL THIOVIT	sulfur		Elemental sulfur
SPORACOL	drazoxolon	GANOCIDE, MIL-COL	4-(2-Chlorophenylhydrazono)-3-methyl-5-isoxazolone
SPOTRETE		*See* thiram	
SPRAY-COP		NEUTROCOP, NU-COP	Copper Sulfate monohydrate
SPRING-BAK		*See* nabam	
STREPCEN	streptomycin	AGRI-STREP, RIMOCIDIN, CHEMFORM, AGRIMYCIN-17	2,4-diguanidino-3,5,6-trihydroxycyclohexyl-5-deoxy-2-0-(2-deoxy-23

TABLE 5.1 *(Continued)*

Trade names are listed in capital letters
Common names are listed in lower case

Compound	Common or Trivial Name	Other Trade Names	Chemical Name
STREPCEN (Contd.)			methylamino-alpha-glucopyranosyl)-3-formyl pentantofuranoside
SULFONIMIDE	captafol	DIFOLATAN, DIFOSAN, SANSPOR	cis-N-(1,1,2,2-Tetrachloro-ethylthio)-4-cyclohexene-1,2-dicarboximide
SULFORON		*See* sulfur	
SULFUR	sulfur (brimstone)	CONSUL, COSAN, COSANGO, HEXASUL, MICROFLOTOX, KOLOFOG, KOLOSPRAY, KUMULUS, SOFRIL, SPERLOX, SPERSUL, THIOLUX	elemental sulfur
SULSOL		*See* sulfur	
SUPER MILGO	ethirimol	MILSTEM	5-m-Butyl-2-ethylamino-4-hydroxy-6-methylpyrimidine
SUZU-H		*See* fentin hydroxide	
SYLLIT		*See* dodine	
TACHIGAREN		HYDROXYISOXAZOLE, HYMEXAZOL	3-hydroxy-5-methylisoxazole
TAYTOX		SOL KOP 10, COPPER COUNT-N	Copper Ammonium complex
TBCS-53		BASICOP, COPPER OXYSULFATE, COPPER 5 DUST, TRIBASIC	Basic Copper Sulfate
TPTA		*See* fentinacetate	
TBZ		*See* thiabendazole	
TCMTB		BUSAN-72, BUSAN-30	2-(Thiocyanomethylthio) benzothiazole
TEC	carbamate	AATECK, TECORAM	bis-(dimethyldithiocarbamoyl)-ethylenebisdithiocarbamate

TECTO	thiabendazole	APL-LUSTR-T, MERTECT, TBZ, MYCOZOL, THIBENZOLE	2-(4-Thiazolyl) benzimidazole
TELONE *11*		*See* dichloropropene	
TERMIL		*See* chlorothalonil	
TERM-I-TROL		*See* pentachlorophenol	
TERRACAP 10-10 DUST		Mixture of 10% PCNB and 10% captan	
TERRAZOLE		*See* etridiazol	
TERSAN-LSR		*See* maneb	
TERSAN-SP		*See* chloroneb	
TERSAN 75		*See* thiram	
TERSAN 1991		*See* benomyl	
	thiabendazole	MERTECT, MYCOZOL, APL-LUSTR-T, TBZ, TECTO, THIBENZOLE	2-(4-Thiazolyl) benzimidazole
THIBENZOLE		*See* thiabendazole	
THIOPHAL		*See* folpet	
	thiophanate	CERCOBIN, CHIPCO SPOT KLEEN, ENOVIT, PELT SOL, TOPSIN, 3336 TURF FUNGICIDE	Diethyl 4,4-o-phenylenebis 3-thioallophanate
	thiophanate methyl	CERCOBIN METHYL, FUNGO, CERCOBIN M, ENOVIT SUPER, TOPSIN M, TOPSIN WP, PELT 44, TREVIN	Dimethyl 4,4-O-phenylebis (3-thioallophanate)
THIOTEX		*See* thiram	
THIRASAN		*See* thiram	
THYLATE		*See* thiram	
TIRAMPA		*See* thiram	
TMTD		*See* thiram	
THYNON		*See* dithianon	
TILCAREX		*See* PCNB	
TILLANTOX		*See* benzuinox	
TOBAZ		Mixture of 25% TBZ, 1.5% malachite green and 0.7% Auramine	
	tolyfluanid	EUPAREN-M	N-dichlorofluoromethylthio-N,N dimethyl-N′-(4-tolyl) sulfamide
TOPANE		DOWICIDE A, SOPP	Sodium-o-Phenyl Phenate
TOPSIN		*See* thiophanate	

TABLE 5.1 *(Continued)*

Trade names are listed in capital letters
Common names are listed in lower case

Compound	Common or Trivial Name	Other Trade Names	Chemical Name
TOPSIN WP METHYL		*See* thiophanate methyl	
TORSITE		DOWICIDE 1, ORTHOXENOL	o-Phenyl Phenol
TPTH		*See* fentin hydroxide	
TPTOH		*See* fentin hydroxide	
TRAMETAN		*See* thiram	
TREVIN		*See* thiophanate methyl	
TRIASYN		*See* anilazine	
TRI-BASIC		*See* Basic Copper Sulfate	
TRICARBAMIX		Mixture of 45% zineb, 15% maneb and 15% ferbam	
TRICARBAMIX FE		Mixture of 15% ziram, 45% ferbam and 15% maneb	
TRICARBAMIX-Z		*See* ziram	
	tridemorph	CALIXIN	N-Tridecyl-2,6,dimethyl-morpholine
TRI-CLOR		*See* chloropicrin	
	tricyclazole	E.L.291	5-methyl-1,2,4-triazole (3,4-b) benzothiazole
TRIFOCIDE		*See* DNOC	
TRIFUNGOL		*See* ferbam	
TRIMANGOL		*See* maneb	
TRIMANOC		*See* maneb	
TRI-PCNB		*See* PCNB	
TRIPOMOL		*See* thiram	
TRISCABOL		*See* ziram	
TRITOFTEROL		*See* zineb	
TRITISAN		*See* PCNB	
TRUBAN		*See* etridiazol	
TSERENOX		See benquinox	
TSITREX		*See* dodine	
TUADS			
	thiram	AATECK, ARASAN, FER-MIDE, FERNACOL, FER-	Tetramethylthiuram disul-fide

		NASAN, HEXATHIR, NOMERSAN, POLYRAM, POMASOL, TERSAN 75, THYLATE, THIOTOX, THIRASAN, TIRAMPA, TUADS	
TUBOTHANE		*See* maneb	
TUBOTIN		*See* fentinacetate	
3336 TURF FUNGICIDE		*See* thiophanate	
TUSCOPPER	copper naphthenate	COPPER CURE, COPPERNATE, COP-R-NAP, CUPRINOL, WITTOX C	cupric cyclopentanecarboxylate
TUTANE		BUTAFUME, FRUCOTE, 2-AB, DECCOTANE	2-AMINOBUTANE
TUZET	Mixture of 20% URBACIDE, 20% ziram and 40% thiram		
UNDONKOR	acetamide compound	CECA	N-(beta-Cyanoethyl) monochloroacetamide
UNICROP		*See* maneb	
URBACIDE		MONZET, URBACID	Methylarsenic-dimethyldithiocarbamate
URBASULF		ASOZIN, MAS, RHIZOCTOL	Methylarsenic sulfide
VALIDICIN			complex antibiotic
VALIDAMYCIN			complex antibiotic
VANCIDE		*See* maneb	
VANCIDE FE 95		*See* ferbam	
VANCIDE MZ 96		*See* ziram	
VANCIDE TM 95		*See* thiram	
VANCIDE ZINEB 85		*See* zineb	
VI-CAD	cadmium compound	CADDY	Cadmium chloride
VIRICUIVRE	basic copper chloride		Copper oxychloride
VITAVAX		*See* carboxin	
VONDCAPTAN		*See* captan	
VONDODINE		*See* dodine	
VORONIT		*See* fuberidazole	
VORONIT SPECIAL		*See* fuberidazole	
WITTOX C	copper naphthenate		Cupric cyclopentanecarboxylate
Z-C SPRAY		*See* ziram	
ZEBTOX		*See* zineb	
ZERLATE		*See* ziram	
ZIDAN		*See* zineb	

TABLE 5.1 *(Continued)*

Trade names are listed in capital letters
Common names are listed in lower case

Compound	Common or Trivial Name	Other Trade Names	Chemical Name
ZIMATE		*See* zineb	
ZINKCARBAMATE		*See* ziram	
ZINC COPOSIL			Basic Cupric Zinc Sulfate Complex
ZINC COPRO			
ZINCMATE		*See* ziram	
	zineb	ASPOR, CHEM ZINEB, DIPHOR, DITHANE Z-78, HEXATHANE, KYPSIN, LONACOL, PARZATE, PARZATE C, POLYRAM, POMOSOL-Z, TIEZENE, TRITOFTOROL, ZEBTOX, ZIDAN, ZINEB 75 WP, ZINOSAN, ZIMATE, VANCIDE-ZINEB 85	Zinc ethylene bisdithiocarbamate
	ziram	COROZATE, CUMAN, FUKLASIN, FUKLASIN ULRA, HEXAZIR, MEZENE, MILBAM, POMARSOL Z FORTE, PRODARANTRISCABOL, TRICARBAMIX-Z, VANCIDE MZ-96, ZERLATE, Z-C SPRAY, ZINCMATE, ZIRBERK, ZIRIDE, ZITOX, ZINCARBAMATE	Zinc dimethyldithiocarbamate
ZIRBERK		*See* ziram	
ZIRIDE		*See* ziram	
ZITOX		*See* ziram	

6

Protective Fungicides for Plant Disease Control

VEGETABLE DISEASE CONTROL

Production of vegetables is a major segment of commercial agriculture in the U.S.A., and is an important part of the home garden project. Americans have learned that vegetables picked fresh from the backyard vegetable patch are a necessary and welcome addition to the daily family diet.

The production of high quality vegetables is big business in New York state, with approximately 185,000 acres devoted to the industry. In this area the vegetable production from approximately 95,000 acres is grown for the freezing and canning industry, and the yield from the remaining 88,000 acres goes to the fresh vegetable market in New York and adjoining areas. Zaumeyer and Thomas (1958) have estimated that vegetable diseases may cost the American farmer $15 to $20 million annually in some seasons.

In the watermelon-producing states of Georgia, Florida, Texas, North and South Carolina, Virginia, Mississippi, and California, and the muskmelon-producing regions of Arizona, California, Colorado, Texas, Michigan, Indiana, Illinois, Maryland, New Jersey, and Delaware, a variety of plant diseases constantly threatens the production of high quality produce.

The annual onion crop in the U.S.A. averages slightly more than 45–50 million 50-pound sacks grown on approximately 117,000 acres in Texas, New York, Michigan, Colorado, California, Oregon, Minnesota, Wisconsin, Idaho, Indiana, and Illinois. Commercial onion growers annually must invest heavily in fungicides for the prevention of destructive onion diseases.

Extensive production of lettuce, tomatoes, radishes, and other vegetables is derived from the winter garden areas of Florida, California, and the Rio Grande Valley of Texas. These are merely a few examples of the importance of commercial vegetable production in the U.S.A.

Vegetables, like other agricultural crops, have their share of destructive diseases that may reduce yields and lower quality. This is especially true in the warmer vegetable producing areas and under irrigation systems. In commercial vegetable production and in the home vegetable garden a knowledge of plant diseases and their control is fundamental for the prevention of serious loss and lowered quality. A knowledge of vegetable diseases and methods for their effective control will serve as insurance against loss of time and money.

As has been previously discussed, the use of chemicals for the prevention of vegetable diseases is only one weapon in our arsenal of defense against these problems. Fungicides for vegetable disease control, when used in combination with other recommended practices, will significantly reduce losses and inconvenience caused by vegetable diseases.

The fungicides used for the prevention of vegetable diseases are summarized in Table 6.1. Specific suggestions for the control of some of the common vegetable diseases are outlined in Table 6.2. Fungicides registered for use on vegetables and the "cut-off dates" to avoid unlawful residues at harvest time are summarized in Table 6.3. See Plates 1-3 for illustrated examples of vegetable diseases.

FRUIT DISEASE CONTROL

Chemical Control of Fruit Diseases in the Home Fruit Planting

Growing fruit at home can be a source of pleasure as well as providing a readily available supply of fresh fruit for home consumption. There is always a real sense of accomplishment in serving to the dinner guest apple, peach, or cherry pie baked from fruit grown in the backyard orchard. Ice cream always tastes especially good when it is topped-off with home-grown fresh strawberries or raspberries. The liquid refreshment that can be brewed from grapes from the home grape arbor is a source of pleasure to some home gardeners, and is a reward for many hours of summer labor, even though the same product is readily available in most supermarkets.

However, destructive diseases have to be accepted as part of the problem of growing fruit at home. While the severity of these problems will vary from year to year depending on the weather, diseases like apple scab, raspberry anthracnose, grape black rot, strawberry leaf spot, and Botrytis fruit rot must be controlled if the home fruit production project is to succeed.

TABLE 6.1. REGISTERED FUNGICIDES FOR VEGETABLE DISEASE CONTROL

Common Name	Trade Name(s)	Chemical Name	Formulations	Approved Uses
Anilazine	Dyrene Kemate	4, 6-Dichloro-N-(2-chlorophenyl)-1, 3, 5 triazin-2-amine	50% WP	Broad spectrum protective fungicide effective against many foliage diseases. Registered for celery, cucumber, muskmelon, onion, potato, pumpkin, squash, tomato and watermelon
Benomyl	Benlate	Methyl 1-(butylcarbamoyl)-2-benzimidazolecarbamate	50% WP	Protective and eradicant fungicide with systemic properties. Registered for beans (dry, lima and snap), celery, cucumber, muskmelon, squash and watermelon
Captafol	Difolatan	cis-N-(1, 1, 2, 2-Tetrachloroethylthio)-4-cyclohexene-1, 2-dicarboximide	80% WP 4 lb/gal. flowable	Broad spectrum, persistent protective fungicide. Registered for use on cucumber, muskmelon, potato, tomato and watermelon
Captan	Captan Orthocide	N-(Trichloromethylthio)-4-cyclohexene-1, 2-dicarboximide	50% and 80% WP 4 lb/gal. flowable 3.5-75% dusts	Protective and eradicative fungicide for seed treatment soil drench and foliar fungicide. Registered for use on beans, beets, carrots, celery, cucumber, eggplant, lettuce, muskmelon, onions, peppers, potatoes, pumpkins, rhubarb, spinach, squash, sweet corn, tomato and watermelon
Chlorothalonil	Bravo Daconil	Tetrachloroisophthalonitrile	75% WP 6 lb/gal. flowable 20% exothermic powder	Broad spectrum fungicide registered for use on bean (snap), broccoli, cabbage, carrots, cauliflower, celery, cucumber,

TABLE 6.1 *(Continued)*

Common Name	Trade Name(s)	Chemical Name	Formulations	Approved Uses
				muskmelon, potato, pumpkin, squash, sweet corn, tomato and watermelon
Dicloran	Botran Allisan	2, 6-Dichloro-4-nitroaniline	50% and 75% WP 4-15% dusts	Used as a soil drench, foliar fungicide and postharvest treatment. Registered for use on snap beans, celery, lettuce, onion, potato, greenhouse rhubarb, cucumber and tomatoes
Dinocap	Karathane Mildex Capryl	Mixture of 2, 4-Dinitro-6-octylphenyl crotonate and 2, 6-Dinitro-4-octylphenyl crotonate	25% WP 4% emulsifiable concentrate	Specific for the control of powdery mildew. Registered for use on cucumber, muskmelon, pumpkin, squash and watermelon
Fentin hydroxide	Du Ter Fenolovo	Triphenyltin hydroxide	20% and 47.5% WP	Organic tin compound used as a foliar fungicide as a protective and curative treatment. Registered for use on potatoes only
Fixed Coppers	Tribasic Kocide C.O.C.S., etc.	Basic sulfates, oxychlorides, and oxides of metallic copper	53-83% WP	Used for the control of bacterial diseases and certain leafspot problems. Registered for use on most vegetables except collards, mustard, parsley, radishes, rhubarb, sweet corn, sweet potatoes and turnips
Folpet	Phaltan	N-(Trichloromethylthio) phthalimide	50% and 75% WP 5% and 10% dusts	Analog of captan. Registered for use on celery, cucumber, lettuce, muskmelon, onion, pump-

				kin, squash, tomato and watermelon
Mancozeb	Dithane M-45 Manzate 200	Maneb + zinc ion	80% WP 4, 8-15% dusts	Broad spectrum fungicide for vegetables. Registered for use on asparagus, beans, carrot, celery, cucumber, muskmelon, onion, potato, squash, sweet corn, tomato and watermelon
Maneb	Dithane M-22 Maneb 80 Manzate	Manganous ethylenebis-dithiocarbamate	70% and 80% WP 1-20% dusts 4 lb/gal. flowable	Foliage protectant. Registered for use on asparagus, beans, broccoli, Brussels sprouts, cabbage, carrots, cauliflower, celery, collard, cucumber, eggplant, endive, kale, muskmelon, onion, parsley, potato, pumpkin, radish, rhubarb (greenhouse), spinach, squash, sweet corn, tomato and watermelon
Metiram	Polyram	Polyethylene carbamate	80% WP 7% dust	Effective against several diseases but will not control powdery mildew. Registered for use on bean, celery, cucumber, muskmelon, potato seed pieces and tomato
PCNB	Terraclor	Pentachloronitrobenzene	35-75% WP 4-99% dusts 75-63% paste	Used primarily as a soil fungicide and seed disinfectant. Registered for use on cabbage, peppers, potatoes, snap beans and tomatoes
Thiram	Arasan Thylate, etc.	Tetramethylthiuram disulfide	90% WP 60% seed protectant 1-75% dusts	Used as a seed protectant for prevention of seed decay, seedling blights and damping-off

TABLE 6.1 *(Continued)*

Common Name	Trade Name(s)	Chemical Name	Formulations	Approved Uses
Zineb	Dithane Z-78 Parzate Polyram-Z	Zinc ethylenebis-dithiocarbamate	75% WP 4 lb/gal. flowable 3-15% dusts	Foliar protectant fungicide. Registered for use on asparagus, bean, beets, broccoli, Brussels sprouts, cabbage, carrot, cauliflower, celery, collard, cucumber, eggplant, endive, kale, lettuce, muskmelon, onion, pea, pepper, pumpkin, radish, spinach, squash, tomato, turnip and watermelon

WP = wettable powder. E.C. = emulsifiable concentrate. F = flowable.

TABLE 6.2. CHEMICAL CONTROL OF VEGETABLE DISEASES

Crop	Disease	Symptoms	Suggested Control
Asparagus	Rust	Orange-red to black pustules on leaves and stems	Use resistant varieties. Apply post-harvest sprays of maneb 80% WP 2 lb/acre (2 tbsp/gal.) or mancozeb 80% WP 2 lb/acre (2 tbsp/gal.) or Polyram 80% WP 2 lb/acre (2 tbsp/gal.)
Beans (Snap and Lima)	Anthracnose	Small angular lesions on leaves and small brown spots on pods which later enlarge to circular, black sunken spots with purple borders	Use western-grown certified seed. Spray at weekly intervals with maneb 80% WP 2 lb/acre (2 tbsp/gal.) or mancozeb 80% WP 2 lb/acre (2 tbsp/gal.) or zineb 75% WP 2 lb/acre (2 tbsp/gal.)
	Bacterial Blight	Leaves have irregular, watery green patches later turning black	Use disease-free certified western seed. Follow a 2-year rotation. Spray with Fixed copper at manufacturer's direction at 7-day intervals or as needed
	Rust	Small reddish-brown pustules on underside of leaves and occasionally on pods	At first sign of disease apply Bravo 6F.3 pt/acre (3 tbsp/gal.) or maneb 80% WP 2 lb/acre (2 tbsp/gal.) or zineb 75% WP 2 lb/acre (2 tbsp/gal.)
	Sclerotinia White Mold	Young plants have watery soft lesions near cotyledon node	Plant in well drained soils not recently planted to beans, potatoes,

TABLE 6.2 *(Continued)*

Crop	Disease	Symptoms	Suggested Control
	Botrytis Gray Mold	with white mat-like mycelium	tomatoes, lettuce or cabbage. Spray at green-bud stage with Benlate 50% WP 1 lb/acre (5 tsp/gal.). Repeat at 7-10 day intervals if disease continues
Beets	Cercospora leaf spot	Circular brown or gray spots with reddish borders on leaves. Dead tissue eventually drops out	Follow rotation. When disease first appears, spray with zineb 75% WP 2 lb/acre (2 tbsp/gal.) or Fixed copper 1-2 lb copper/acre (2 tbsp/gal.) If disease persists, repeat at 7-10 day intervals
Cabbage, Cauliflower, Broccoli and Brussels sprouts	Blackleg (Phoma) and Black Rot	Dark, sunken circular lesions on stems and leaves. Stems often girdled near soil line	Use western-grown certified seed and hot-water treated seed. Follow 4-year rotation free of cruciferous crops
	Downy mildew	White downy mold on leaves which later shrivel and die	Use hot-water treated seed or certified, disease-free seed. Spray at 7-10 day intervals with Bravo 6F 1½ pt/acre (2 tbsp/gal.) or maneb 80% WP 2 lb/acre (2 tbsp/gal.) or mancozeb 80% WP 2 lb/acre (2 tbsp/gal.)
	Alternaria	Leaf spots 2-3 in. in diameter. Later developing concentric circles and take on a black sooty color	Add maneb 80% WP at 2 lb/acre to midseason insecticide sprays

Carrots	Leaf spots	Dark spots on leaves and petioles, resulting in yellowing and death of leaves	Follow 2-3 year rotation. Apply 3-5 sprays at 7-10 day intervals of Bravo 75% WP 2 lb/acre (2 tbsp/gal.) or zineb 75% WP 3 lb/acre (2 tbsp/gal.) or Fixed Copper at manufacturer's direction
Celery	Fusarium yellows or wilt	Yellowing of leaves. Plants stunted. Stalks brittle and bitter to taste	Plant wilt-resistant varieties and follow 4-5 year rotation
	Early blight	Irregular, velvety, ash gray dead areas on leaves and petioles	Treat seed with thiram. Apply 3-5 sprays at 10-day intervals starting when symptoms first appear. Bravo 75% WP 2 lb/acre (2 tbsp/gal.) or mancozeb 80% WP 3 lb/acre (3 tbsp/gal.) or Benlate 50% WP ½ lb/acre (2 tsp/gal.)
Cucurbits (cucumber, melons, pumpkin and squash)	Angular leaf spot	Water-soaked, angular spots on leaves which later dry up and fall out	Use resistant varieties and purchase disease-free transplants. Spray with fixed copper 1-1½ lb/acre (actual copper) (2 tbsp/gal.) at 5-7 day intervals until mid-harvest if disease persists
	Anthracnose and Early Blight	Brown spots on leaves. Sunken circular cankers with pink centers and brown margins	Use disease-free seed or transplants. Spray at 7-10 day intervals starting at 2-leaf stage using maneb 80% WP 2 lb/acre (2 tbsp/gal.) or Benlate 50% WP ¼-½ lb/acre (2 tsp/gal.)

TABLE 6.2 *(Continued)*

Crop	Disease	Symptoms	Suggested Control
			or Bravo 6F 1½-2 pt/acre (2 tbsp/gal.) or Difolatan 4F 4 pt/acre
	Bacterial Wilt	Rapid wilting of plants with white sticky ooze from cut stems	Apply insecticide to control insects that transmit causal bacteria
	Powdery mildew	White powdery mold on leaves and stems	Spray with Benlate 50% WP ¼-½ lb/acre (2 tsp/gal.) or Karathane 22% WP 8 oz/100 gal. (1 tsp/gal.) Reapply at 7-10 day intervals starting as soon as mildew appears
	Downy mildew	Yellow angular spots on upper leaf surfaces with white-purplish mold on lower leaf surfaces	After runners are formed, apply at 7-10 day intervals Bravo 75% WP 2 lb/acre (2 tbsp/gal.) or mancozeb 80% WP 2 lb/acre (2 tbsp/gal.)
	Scab	Sunken spots with oozing sap and greenish-black mold on fruit, followed by fruit rot	Plant resistant varieties and apply at 7-10 day intervals Bravo 70% WP 2 lb/acre (2 tbsp/gal.) or Bravo 6F 2 pt/acre or mancozeb 80% WP 2-3 lb/acre (2 tbsp/gal.)

Eggplant	Anthracnose and Phomopsis fruit rot	Circular gray-brown areas on leaves. Light brown, circular and depressed spots on fruits with black fruiting bodies. Dry rot cankers on stems of older plants	Use hot-water treated seed—122°F for 25 min. Spray at 10-day intervals with Captan 50% WP 2 lb/acre (2 tbsp/gal.) or maneb 80% WP 2 lb/acre (2 tbsp/gal.) or zineb 75% WP 2 lb/acre (1½ tbsp/gal.)
	Leaf Spots	Several types on leaves	Same as anthracnose
Lettuce	Bottom rot and drop	Water-soaked spots on stem at soil-line spreading vertically. White mass of mycelium with embedded black sclerotia on infected areas	No effective fungicide. Follow 3-year crop rotation
	Botrytis Gray Mold	Water-soaked discolored areas on lower leaves and stems. Ultimately entire plant becomes covered with a gray mold and decays	After thinning when plants have recovered, spray with Botran 75% WP 2 lb/acre (2 tbsp/gal.) or maneb 80% WP 2 lb/acre (2 tbsp/gal.)
	Downy mildew	Light areas develop at the margins of infected leaves, followed by the appearance of a white mold on the lower leaf surfaces. Infected plants turn yellow and die	Spray with zineb 75% WP 2 lb/acre (2 tbsp/gal.) or maneb 80% WP 2 lb/acre (2 tbsp/gal.)
Onion	Downy mildew	Pale green, oval to elongated lesions on leaves. In wet cool weather a purple mold may develop on infected leaves and onion tops die	As soon as disease first appears, spray at 10-day intervals with maneb 80% WP 2 lb/acre (2 tbsp/gal.) or mancozeb 80% WP 2 lb/acre

TABLE 6.2 *(Continued)*

Crop	Disease	Symptoms	Suggested Control
			or Bravo 6F 1½-3 pt/acre or Difolatan 4F 2½ pt/acre or Dyrene 50% WP 3 lb/acre
	Purple Blotch	Reddish-purple spots on onion tops surrounded by a yellow zone	Same as Downy mildew
	Smut	Leaves and bulbs of young plants covered with a mass of black powdery smut spores	Dust or pellet seed with thiram or Captan at manufacturer's direction. At planting time, apply 1-2 lb of thiram per 14,500 linear feet of furrow. Avoid smut-infested soil
Peppers	Anthracnose and fruit rots	Tan or gray sunken lesions on fruit with pink fruiting bodies in lesions	Spray at 7-10 day intervals through harvest with maneb 80% WP 2 lb/acre or zineb 75% WP 2 lb/acre
	Bacterial Spot	Dark brown raised lesions on leaves which turn yellow and fall. Rough, blister-like spots may develop on fruit	Use disease-free seed and apply 1.0-1.6 lb of copper (Fixed copper) per acre through flowering period, e.g., Kocide 101. After fruit set apply maneb 80% WP 2 lb/acre or zineb 75% WP 2 lb/acre
	Blossom-end rot	See Tomato	
	Cercospora Leaf Spot	Small, circular, water-soaked spots on leaves enlarging to gray spots with reddish-brown	Same as anthracnose

		margins. Centers of spots usually fall out	
Potato (Irish)	Early Blight	Circular brown spots with concentric rings giving a target-spot effect	When plants are 8-12 in. high start fungicide treatments and repeat at 7-10 day intervals maneb 80% WP 2 lb/acre or mancozeb 80% WP 2 lb/acre or Difolatan 4F 3 pt/acre or Du Ter 47.5% WP 5-10 oz/acre or Bravo 6F 1-1½ pt/acre or Polyram 80% WP 2 lb/acre
	Late Blight	Brownish to purplish irregular lesions on leaves. A white mold may develop on the lower leaf surface in wet weather. Tubers may become infected and rot	Same as Early Blight
	Scab	Tubers with raised, corky brown areas which are frequently circular and sometimes pitted	Plant scab-free seed pieces in non-infested soils. Use resistant varieties and maintain soil at pH 5.2 or lower
	Rhizoctonia (Black Scurf)	Top growth stunted with small aerial tubers. Brown cankers usually occur on stems at soil level	Plant only certified, disease-free seed pieces. Infested soil may be treated with PCNB (Terraclor 75% WP) according to manufacturer's directions
Spinach	Downy mildew	Pale yellow spots on leaves covered on the underside with a blue-gray mold. Plants light in color and stunted	During wet cool weather spray at 7-10 day intervals with maneb 80% WP 1½ lb/acre or zineb 75% WP 2 lb/acre. Plant resistant varieties

TABLE 6.2 *(Continued)*

Crop	Disease	Symptoms	Suggested Control
	White Rust	White spots on leaves	Spray with fungicides suggested for downy mildew
Sweet Corn	Leaf Blight	Long, elliptical or small, somewhat blunt lesions on leaf blades and ear sheaths	The use of fungicides is generally not profitable. Plant varieties with Normal (N) cytoplasm
Sweet Potato	Black Rot	Black cankers on underground stems. Circular, shallow brown spots on tubers which are bitter to the taste	Use disease-free bedding stock. Follow crop rotation. Dip seed in thiram 75% WP 1 lb/5 gal. of water
	Scurf	Brown, irregular, superficial discolored spots on the root	Same as Black rot
	Pox	Infected plants yellow and stunted with many lateral roots destroyed. Brown sunken spots on tubers	Use disease-free seed. Maintain soil below pH 5.2. Follow crop rotation
Tomato	Early blight	Dark brown, circular spots with concentric rings on leaves. Lesions on fruit are dark brown or black. Fruit rots	Use disease-free transplants. Follow crop rotation. Apply fungicides throughout the season at 7-10 day intervals using: maneb 80% WP 2-3 lb/acre or Bravo 6F 1½-3 pt/acre or Difolatan 4F 4 pt/acre or Dyrene 50% WP 2-3 lb/acre or Polyram 80% WP 3 lb/acre or mancozeb 80% WP 2 lb/acre

Late blight	Water-soaked, brown to purplish irregular spots on leaves. Undersides of infected leaves develop a white mold under moist conditions. Fruit rots	Same as Early Blight
Gray Leaf Spot	Small brown spots with gray centers on infected leaves starting on oldest leaves. Leaves fall prematurely	Apply fungicides suggested for early blight
Septoria Leaf Spot	Small, water-soaked circular spots on leaves and stems. Spots with gray centers and dark brown border. Small black fruiting bodies (pycnidia) develop in lesions	Same as early blight
Anthracnose	Water-soaked, sunken circular spots on fruit, usually with concentric rings. Fruit rot develops on ripe fruit	Starting when first fruits develop and repeating at 10-day intervals throughout season, apply Folpet 50% WP 2 lb/acre or Bravo 75% WP 2 lb/acre or mancozeb 80% WP 2 lb/acre or zineb 75% WP 2 lb/acre

WP = wettable powder. F = flowable formulation. 1 lb/acre = approximately 1 tbsp/gal. water. tsp = teaspoonful.

TABLE 6.3. TOLERANCE RESTRICTIONS FOR VEGETABLE FUNGICIDES AND THEIR SUGGESTED "CUT-OFF DATES"

No number after X = No time limitation. Numbers after X = days between last application and harvest.

Vegetable Crop	Benlate	Botran	Bravo	Captan	Difolatan	Du Ter	Dyrene	Fixed Copper	Karathane	Maneb	Mancozeb	Polyram	Terraclor	Thiram	Zineb
Asparagus				X						X	X	X			X
Beans (Lima)	X.28			X				X		X.4			X		X.7
Beans Snap	X.14	X.2	X.7	X				X		X.4			X	X	X.7
Beets				X				X						X	X
Cabbage			X	X				X		X				X	X
Cantaloupe	X		X	X	X		X	X	X.7	X.5	X.5	X.5		X	X.5
Carrots		X	X	X				X		X	X.7			X	X.7
Collards				X						X				X	X
Celery	X.7	X.7	X.7	X			X			X	X.14				X.14
Cucumber	X	X	X	X	X		X	X	X.7	X.5	X.5	X.5		X	X.5
Eggplant				X				X		X				X	X
Endive										X					X
Lettuce		X.14		X				X		X				X	X.10
Mustard				X						X				X	X.10
Onions		X	X	X	X		X	X		X	X.7				X.7
Parsley															X
Peppers				X				X		X			X	X	X
Potatoes		X.14	X	X	X	X.7	X	X		X	X	X	X		X

Pumpkins			X	X			X	X	X.7	X				X	X
Radishes								X						X	X
Rhubarb		X		X						X					
Spinach				X			X	X		X				X	X.10
Squash	X		X	X			X	X	X.7	X.5	X.7			X	X.5
Sweet Corn			X.14	X						X	X.7	X.1		X	X
Sweet Potato		X												X	
Tomatoes	X	X	X	X	X		X	X		X.5	X.5	X.5	X	X	X.5
Turnips				X						X				X	X
Watermelons	X		X	X	X		X	X	X.7	X.5	X.5			X	X.5

When using any of the above fungicides, check the label on the container and observe all precautions. A blank space in the above table indicates that the fungicide is not registered for the crop indicated.

Most home fruit plantings can be relatively free from plant disease problems but it requires a little extra care in planning the summer garden program. Unlike insect problems, the symptoms of the common fruit diseases generally are not apparent until two to three weeks after they have been initiated. By this time it is usually too late to prevent the damage and disappointment they can cause. There are several plant diseases that may attack fruit crops throughout the growing season. The common diseases of fruit are specific for each fruit crop; the disease problems that can occur on apples do not affect peaches, and strawberry diseases do not attack grapes. There is a specific cause for each of the plant disease problems in the backyard fruit planting. Some typical diseases of peach trees are shown in Plate 4, and those of apple trees are shown in Plate 5.

There is no single practice or miracle cure that will control all of the plant disease problems of fruit in the home garden.

A combination of good horticultural practices and the judicious and timely use of fungicides will prevent the unwelcome occurrence of unsightly and destructive fruit diseases. Fruit trees and small fruit plantings must be given a chance to resist attack by plant disease pathogens. Most fruits rarely achieve the appearance of seed catalogs without attention to annual pest control programs. This means selecting the proper planting site, the use of locally adapted varieties, the application of prescribed fertilizers, correct pruning, and other horticultural suggestions. One of the most important practices in the home fruit planting is regular sanitation. This consists of an annual clean-up, and removing dead and diseased branches from fruit trees and old fruiting canes from berry crops such as raspberries and grapes. Fallen leaves also should be raked up and destroyed in the fall since these can be the source of spring disease infections such as apple scab.

In most seasons the use of chemicals for fruit disease control is a necessary supplement to good horticultural practices. There is no one fungicide that will control all of the diseases of fruit crops; neither will one or two spray applications made on free weekends or at other convenient times suffice to control fruit disease problems. To be effective, fungicides for disease prevention must be applied at the right time, in the right way, using the recommended dosage, and continuing during periods when disease infections can take place. In recent years most garden supply stores, nurseries, and some department stores have offered the home fruit grower "GENERAL PURPOSE" or "MULTIPURPOSE FUNGICIDE-INSECTICIDE" formulations containing at least one fungicide and one insecticide. The use of these commercial general purpose fruit spray formulations greatly simplifies the problem of disease and insect control in the home fruit planting. Homeowners have the choice of using a multipurpose pesticide or they may use specific fungicides and insecticides to control specific diseases and insect pests. The choice of which pesticide to use depends on the amount of time, energy, and money the home grower wishes to expend.

Several commercially available "multipurpose" or "general purpose" fungicide-insecticide fruit spray formulations are available at most farm supply stores, garden shops, and nurseries. Examples of these are given below:

Ortho Orthocide Garden Fungicide. Chevron Chemical Co., San Francisco, California
Pop-in Fruit Tree Spray. Encap Products Co., Mt. Prospect, Illinois
Henry Fields's MultiPurpose Spray. Henry Field Seed & Nursery Co., Shenandoah, Iowa
Farm Bureau One Package Fruit Spray

A homemade general purpose fruit spray may be prepared by buying the pesticides separately and mixing them according to the proportions summarized below:

Material	Amount Required to Make		
	1 gal.	5 gal.	20 gal.
Captan 50% WP plus	2 TBS	1 cup	¾ lb
Methoxychlor 50% WP plus	2 TBS	1 cup	¾ lb
Malathion 25% WP	2 TBS	¾ cup	½ lb

WP = wettable powder. TBS = level tablespoonful.

Commercially available general purpose fruit spray formulations have the advantage of making it unnecessary for the home fruit grower to identify the disease or insect problem to be controlled. These formulations do not provide the high degree of control given by specific fungicides or insecticides. However, in small fruit plantings they will provide a reasonable degree of control for the common fruit diseases and insect pests with minimal effort.

For the home fruit grower who wishes to use more effective fungicides for specific fruit diseases the following products are registered and approved for use on fruit crops:

Benomyl is sold under the trade name Benlate as a 50% wettable powder. It is a broad spectrum fungicide with protective and eradicant properties, and can effectively control apple scab, powdery mildew, sooty blotch, and flyspeck diseases of apples. It also prevents storage rots and is registered for pear disease control. Benomyl is registered for use on peaches, nectarines, apricots, cherries, prunes, and plums for the prevention of brown rot blossom blight and fruit rot, powdery mildew, cherry leaf spot, and peach scab. It may also be used for the control of blueberry, cranberry, grape, and strawberry diseases.

Captan is marketed as a 50% and 80% wettable powder and as a 3.5 to 70% dust under trade names such as Orthocide 50 and Captan 50W, and is still one of the most effective fruit fungicides. It will give excellent control of apple scab and Brook's Fruit spot, and is reasonably effective for the control of Botryosphaeria fruit rot, bitter rot, black rot, flyspeck, sooty blotch, and black pox diseases of apples. It will not prevent apple cedar-rust, powdery mildew, or fireblight of apples. It is also effective against peach and plum brown rot, cherry leafspot, strawberry leafspot and fruit rots, and raspberry anthracnose and Botrytis fruit rot. It is also suggested for the prevention of Black rot, Dead arm disease, and downy mildew on grapes.

Difolatan is a relatively new fungicide available as an 80% wettable powder and as a 4 lb/gal. flowable formulation. It has been widely used for the control of primary infections of apple scab from the delayed dormant stage to petal fall stage. It has become popular because of the Single Application Technic (S.A.T.) which eliminates the need for the conventional series of sprays previously used for primary apple scab control. It is not effective for secondary apple scab, summer diseases of apples, or powdery mildew.

Dikar is a mixture of 74% mancozeb (Dithane M-45) and Karathane (4.7%), registered for use only on apples. It gives good control of apple scab, apple cedar rust, quince rust, and apple fruit rots, and also is a suppressant for mites.

Dodine is the generic name of the fungicide sold under the trade name Cyprex. It is available as a 60% and 80% wettable powder, as a 1–4% dust, and as a 20% and 25% liquid. It is used primarily for the control of apple scab, and has some eradicant properties. It also gives excellent control of cherry leaf spot.

Ferbam was the first organic fungicide to be approved for use on fruit. It is marketed as a 76% and 95% wettable powder and as a 6–25% dust. It is a black powder distributed under trade names such as Fermate, Niacide, Ferberk, and Vancide-FE95. It is suggested for the control of cedar apple rust, quince rust, grape black rot, raspberry anthracnose, and peach leaf curl. It is not considered to be safe on peaches when applied as a foliar spray. It is not particularly effective for the control of apple scab, sooty blotch, cherry leafspot, or brown rot of peaches, plums, and other stone fruits.

Folpet is an analog of Captan sold under the trade name Phaltan as a 50% and 75% wettable powder and as a 5% and 10% dust. Phaltan is one of the most effective fungicides for the control of black rot of grapes and will control both powdery and downy mildew on grapes. It may be used for leafspot and brown rot of sour cherries but is injurious to peaches. It is very effective for sooty blotch, flyspeck and summer rots of apples. If used for the control of apple scab it should not be used in the cover sprays.

Karathane is available as a 25% wettable powder and a 48% liquid formulation. It is a specific for powdery mildew control and will not control other fruit diseases.

Mancozeb is the generic name for a carbamate fungicide sold under the trade names Manzate 200 and Dithane M-45 as an 80% wettable powder. It has been approved for the control of apple scab, apple cedar rust, summer rots of apple, crabapple, pear, and quince, and for black rot, dead arm and downy mildew of grape. It is not registered for use on stone fruits.

Maneb is distributed under the trade names Manzate D and Dithane M-22 Special as an 80% wettable powder formulation containing zinc, and is used for the same purposes as mancozeb.

Polyram is a mixture of dithiocarbamate fungicides and is available as an 80% wettable powder. It controls apple scab, cedar rust, sooty blotch, and flyspeck on apples, but is not effective for the control of powdery mildew.

Thiram is distributed under the trade name Thylate as a 65% wettable powder. Excellent control of strawberry leafspot and fruit rots results from the use of this fungicide. It may also be used for the control of apple scab and apple cedar rust.

Zineb is one of the earlier carbamate fungicides and is sold as a 75% wettable powder under the trade names Dithane Z-78 and Parzate. It is not as effective as some other organic fungicides for the control of apple scab, but is suggested as an additive fungicide for the control of cedar apple rust in pre-bloom and petal fall sprays for apples. When used in combination with Captan it improves control of summer rots and sooty blotch of apples and frogeye leaf spot.

Multipurpose, or general purpose, fruit spray formulations are the most convenient method of controlling fruit diseases in the home fruit planting. The suggested timing schedule using multipurpose fruit spray formulations is outlined in Table 6.4. Home fruit growers who prefer to use the more effective organic fungicides should refer to the information outlined in the suggestions for the control of fruit diseases for the commercial grower.

Chemical Control of Commercial Fruit Crop Diseases

The use of chemicals for the prevention of diseases of commercial fruit crops is one of the few universally accepted examples of the need for chemical controls for destructive plant diseases. In concentrated plant-

TABLE 6.4. SPRAY SCHEDULE FOR THE CONTROL OF FRUIT DISEASES IN THE HOME FRUIT GARDEN

Time to Spray	Apples	Pears	Peaches and Plums	Cherry	Raspberry	Grapes	Strawberries
DELAYED DORMANT	Superior Oil	—	Ferbam	—	Lime-sulfur (2 C/gal.)	—	—
HALF-INCH GREEN When ½ in. of green tissue protrudes from opening buds	Benomyl or Captan	—	—	—	—	—	—
PRE-BLOOM	XX	XX	XX	XX	XX	XX	XX
FULL BLOOM	Captan	—	—	—	—	—	—
PETAL FALL When 75% of petals have fallen	XX	XX	XX	XX	XX	XX	XX
FIRST COVER 7-10 days	XX	XX	XX	XX	—	—	—
ADDITIONAL COVER SPRAYS	XX	XX	XX	XX	—	XX	Captan or Benomyl
MINIMUM NUMBER	6-8	4-6	6-8	6	3	4-5	3-4

XX indicates application of general purpose or multipurpose fruit formulation.

ings of perennial crops, such as apples, pears, peaches, plums, and cherries, diseases can become more destructive than they often are in annual crops such as potatoes, tomatoes and other vegetables. Once diseases become established in the commercial apple or peach orchard they are generally difficult to eliminate and usually require annual applications of fungicides to keep them under commercial control. For this reason, fruit fungicides have received particular attention from research workers and the agricultural chemical industry. Through the years many different fungicides have been developed for fruit disease control.

Of necessity, commercial fruit growers have become highly skilled in the use of chemical tools for fruit disease prevention. Many of the early developments in fungicides were in chemicals for fruit disease control. Thus, liquid lime-sulfur was accidentally discovered to be an effective preventative for peach leaf curl by A.B. Cordley in Oregon in 1908. Lime-sulfur was widely used at that time as a sheep dip for tick control in Australia where it was called "Victoria Lime-and-Sulfur Dip." Previously, in France, Grison, the head gardener of the vegetable houses of Versailles, had adapted lime-sulfur for plant disease control in 1851 and had named it "Grison's Liquid." A.J. Farley of the New Jersey Agricultural Experiment Station developed "Dry-mix Sulfur-lime" for the prevention of diseases of stonefruits in 1922. Colloidal sulfur was introduced by H.C. Young of the Ohio Agricultural Experiment Station in 1924.

The need for the effective and safe use of fungicides for the control of fruit diseases in commercial plantings has been influenced in recent years by the necessity for producing fancy and extra fancy fruit for the consumer market. A high percentage of packout of fancy fruit grades is essential not only for the commercial fruit grower, but the average person will not accept and purchase low quality fruit.

The use of fungicides in commercial fruit plantings is regulated by the EPA (Environmental Protection Agency) which requires that all private applicators of pesticides in the commercial fruit orchard be certified in order to apply "restricted use" pesticides. Certification by the EPA and state agencies requires that all commercial orchardists understand and observe label restrictions on the fungicide formulation labels. Specific safety precautions and information on the environmental consequences of the use and misuse of pesticides must also be observed. Furthermore, fruit fungicides must be approved for use on specific fruit crops by the EPA and State Pesticide Regulatory Agencies. Harvest restrictions and residue tolerance levels have been established for all fruit fungicides as well. Exceeding these restrictions is a violation of federal law, and fruit with excess fungicide residues at harvest time can be legally removed from the consumer market.

In 1966 approximately 160 million pounds of elemental sulfur were used by fruit growers in the United States, mostly for the prevention of apple scab. In recent years more effective fungicides for fruit disease control have been introduced and accepted widely by commercial growers of fruit crops. Among the newer fruit fungicides may be listed Cyprex (Dodine), Karathane (capryl), Botran (dicloran), and Phaltan (folpet). The identities and uses of modern fungicides for fruit disease control are summarized in Table 6.5.

The following summarized information on chemicals for fruit disease control is designed to aid the commercial fruit grower in the selection of chemicals for particular problems of individual fruit crops. Since these practices cannot be "blueprinted" to fit all geographical locations, the reader is urged to consult the Extension Plant Pathologist at the state college of agriculture, or the local county agricultural agent for regional suggestions for fruit disease control.

Cut-off Dates for Fruit Fungicides

The EPA's tolerances for fruit fungicide residues on harvested fruit are expressed as parts per million (PPM) of the active ingredient of the fungicide remaining on the fruit. These tolerances are available from the EPA and state pesticide regulatory agencies.

The commercial fruit grower is unequipped and unable to conduct chemical analyses to determine the PPM of fruit fungicide residue. For this reason, "cut-off dates" were established to help commercial fruit growers avoid violation of residue tolerances of fungicides on harvested fruit. The "cut-off date" is the time (in days) between the final application of the fungicide and the harvest date of the treated fruit. Cut-off dates for the various fruit fungicides have been established by careful research to provide a margin of safety in avoiding unlawful residues of fruit fungicides.

The currently used fruit fungicides are biodegradable, and in most cases deteriorate and become innocuous after 10-21 days' exposure to atmospheric conditions. The "cut-off dates" for the currently used fruit fungicides are summarized in Table 6.10.

Acquired Resistance to Fruit Fungicides

Until recently, acquired resistance to organic fungicides by fungi was uncommon. However, recent observations by plant pathologists in the major fruit-producing areas indicate that there now are strains of some fruit pathogens resistant to some organic fungicides.

TABLE 6.5. FUNGICIDES FOR COMMERCIAL FRUIT DISEASE CONTROL

Common Name	Chemical Name	Trade Name	Formulations	Manufacturer	Diseases Controlled	Suggested Dosage
Benomyl	methyl 1-(butyl carbamoyl)-2-benzimidazole-carbamate	Benlate	50% WP	DuPont	Apple scab, powdery mildew, sooty blotch, flyspeck, storage rots. Peach, apricot, nectarine, cherry, plum and prune Brown rot, Peach scab, powdery mildew, cherry leafspot and post-harvest rots. Also approved for blueberries, cranberries, grapes and strawberries	4-6 oz/100 gal.
Bordeaux mixture	Mixture of copper sulfate and hydrated lime	Bor-Dox Copper Hydro Bordo, Ortho Bordo mixture, etc.	2-3-100 to 8-8-100	Sherwin Williams Co. DuPont Co. Tennessee Corp. Chevron Chemical Co., Ortho Div.	Apple and pear fireblight Peach leaf curl Cherry leaf spot Apple blotch and black rot	2-6-100 10-10-100 or 6-6-100, 1-3-100 8-8-100
Captan	N-trichloro-methylthio-4-cyclohexene-1, 2-dicarboximide	Captan Orthocide	50% WP	Chevron Chem. Co., Ortho Div. Stauffer Chem. Co.	Apple: scab, black rot, Brook's fruit spot, B. ribis, Botrytis blossom end rot. Peach: Brown rot and Rhizopus rot. Plum: Brown Rot and Rhizopus Rot. Cherry: leaf spot and Brown rot. Strawberry: Leaf spot and harvest rot. Raspberry: Anthracnose. Grape: Dead arm, black rot and downy mildew	2 lb/100 gal.
Cycloheximide	Beta-2-(3, 5-dimethyl-2-oxocyclohexyl-2-hydroxyethyl)-glutarimide	Acti-dione	7.7% tablet 0.027% and 0.5% dust, 4.0% LC	Tuco Products, Div. UpJohn Co.	Cherry Leaf spot and powdery mildew	1 PPM
Dichlone	2, 3 -dichloro -1, 4-naphthoquinone	Phygon	50% WP	U.S. Rubber Co.	Apple scab Peach and Plum Brown rot	½ lb/100 gal. ¼ lb/100 gal.
Dicloran	2, 6-dichloro-4-nitroaniline	Botran	50% WP 75% WP 6% dust	Tuco Products, Div UpJohn Co.	Brown rot of stone fruits Grape Botrytis rot	1-1⅓ lb/100 gal. or 30 lb 6% dust per acre

TABLE 6.5 *(Continued)*

Common Name	Chemical Name	Trade Name	Formulations	Manufacturer	Diseases Controlled	Suggested Dosage
Difolatan (captafol)	cis-n (1, 1, 2, 2-tetra-chloro-ethyl)-4-cyclo-hexene-1, 2-dicarboximide	Difolatan	80% WP 4 lb/gal. flowable	Chevron	Primary apple scab Not suggested for secondary scab, powdery mildew	3-5 qt/100 gal.
Dikar	Mixture of mancozeb (Dithane M-45) and dinocap (Karathane)	—	80% WP	Rohm & Haas Co.	Apple scab, powdery mildew, apple and quince rusts and summer apple rots	1½ lb/100 gal.
Dinocap	Dinitro phenyl crotonate	Karathane	24% WP 48% LC	Rohm & Haas Co.	Powdery mildew	¼-½ lb/100 gal.
Dodine	N-dodecyl-guanidine acetate	Cyprex	65% WP 3% dust	Am. Cyanimid Co.	Apple scab Cherry leaf spot	¼-¾ lb/100 gal.
Ferbam	Ferric dimethyldithio-carbamate	Fermate, Ferradow, Coromate, Ferberk, Karbam Black, etc.	76% WP 95% WP	DuPont Co. Union Carbide Niagara Chem. Co. Rohm & Haas Co., etc.	Apple: scab, cedar rust, summer rots, Cherry leaf spot, Grape Black rot and Downy mildew, Peach leaf curl, Currant and gooseberry leaf spot	1-2 lb/100 gal.
Fixed coppers	Basic salts of copper sulfate	C.O.C.S., Tri-Cop, Microgel, Spray-cop, etc.	Wettable powders	Niagara Chem. Div., FMC Corp. Tennessee Corp. DuPont, Harshaw Chem. Co. etc.	Substitute for bordeaux mixture	At manufacturer's direction
Folpet	N-trichloromethyl thio-phthalimide	Phaltan	50% WP	Chevron Chem. Co., Ortho Chem. Div.	Apple: scab, black rot Flyspeck and sooty blotch. Cherry: leaf spot, brown rot and Botrytis rot. Grape: dead arm, downy mildew and powdery mildew	1½ lb/100 gal.
Glyodex	Mixture of dodine 22.5% and glyodin 37.5%	Glyodex	Wettable powder	Union Carbide Chem. Co.	Apple scab	½ lb/100 gal.

Glyodin	2-heptadecyl glyoxalidine acetate	Glyodin	30% LC	Union Carbide Chem. Co.	Apple scab, Brook's fruit spot and sooty blotch	2 pt/100 gal.; or 1 pt in combination with other fungicide
Lime-sulfur	30-32 Baumé calcium polysulfides	Lime-sulfur Orthorix Eau Grison	Liquid concentrate	Chevron Chem. Co. Ortho Div., Dow Chem. Co. Allied Chem. Corp, etc.	Apple scab Peach leaf curl Raspberry anthracnose	2 gal./100 delayed dormant only 5 gal./100 dormant only. 12 gal./100 delayed dormant only
Mancozeb	Coordination product of zinc ion and manganese ethylene bisdithiocarbamate	Dithane M-45 Manzate 200	80% WP	DuPont Rohm & Haas Co.	Apple scab, cedar rusts, black rot, bitter rot, and sooty blotch of apple, crabapple, pear and quince. Also grape black rot, Dead arm, downy mildew. Not registered for stonefruits	1.5-2.5 lb/100 gal.
Maneb	Manganese ethylene bisdithiocarbamate plus zinc	Manzate "D" Dithane M-22 Special	Wettable powder	DuPont Rohm & Haas Co.	Apple scab, cedar rust, black rot, bitter rot, sooty blotch and flyspeck	2 lb/100 or ¼ lb added to other fungicide
Niacide "A"	35% ferbam, 24% manganous dimethyl dithiocarbamate, 6.2% thiram, 1.1% 2, 2-dithiobisbenzothiazole	Niacide	Wettable powder	Niagara Chem. Div., FMC Corp.	Apple scab and cedar rust	2 lb/100 gal.
Niacide "M"	Same as Niacide "A" without ferbam	Niacide	Wettable powder	Niagara Chem. Div., FMC Corp.	Same as Niacide "A"	2 lb/100 gal.
Phybam-S	3% dichlone, 9.5% ferbam	Phybam	50% WP 1-4% dust	Niagara Chem. Div., FMC Corp.	Apple scab, cedar rust and powdery mildew	4 lb/100 gal.
Polyram	Zinc polyethylene thiuram disulfide	Polyram	80% WP 5% dust	Niagara Chem. Div. FMC Corp.	Apple scab, sooty blotch and flyspeck	2 lb/100 gal.
Streptomycin	Streptomycin sulfate	Agri-mycin Agri-strep	15-20% WP	Chas. Pfizer & Co., Merck & Co., Inc.	Apple, pear and quince fireblight	50-100 ppm

TABLE 6.5 *(Continued)*

Common Name	Chemical Name	Trade Name	Formulations	Manufacturer	Diseases Controlled	Suggested Dosage
Sulfur	30-80% elemental sulfur	Flotox, Kolodust, Kolofog, Microsul, Magnetic '70, etc.	Dusts, wettable powders, pastes and fused bentonite sulfurs	Stauffer Chem. Co. Olin Mathieson, etc.	Apple scab and powdery mildew. Powdery mildew on other fruits	5-6 lb/100 gal.
Thiram	Tetramethylthiuram disulfide	Thylate	65% WP 5% dust	DuPont	Apple scab and cedar rust. Strawberry leaf spot and fruit rot	2 lb/100 gal. 4 lb/acre
Zineb	Zinc ethylenebisdithiocarbamate	Dithane Z-78 Parzate	75% WP 4-16% dust	DuPont, Rohm & Haas Co.	Plum Black knot, apple cedar rust, sooty blotch and flyspeck and summer rots	2 lb/100 gal. or 1 lb/100 gal. in combination with other fungicides

Some common spreaders and stickers: DuPont Spreader-Sticker
Filmfast
Nu Film
Ortho Spreader-Sticker
Spread Rite
Triton B. 1956
Triton X-100

WP = wettable powder. LC = liquid concentrate. 1 lb/100 gal. = 1 tbsp/gal. (approximately).

TABLE 6.6. SPECIFIC SUGGESTIONS FOR THE COMMERCIAL CONTROL OF APPLE DISEASES

Growth Stage	Diseases Controlled	Materials and Amounts in 100 gal. and General Comments	
DORMANT: before growth starts in spring and when temperatures are above 45°F	Fireblight (*Erwinia amylovora*), Black Rot (*Physalospora malorum*), Blotch (*Phyllosticta solitaria*)	Copper sulfate (Blue vitriol) 4 lb or 8-8-100 bordeaux mixture plus 3 gal. emulsifiable oil. Do not combine copper sulfate with dormant oil	
	Sepal and leaf scab to pink stage	Difolatan 4F 3 qt/100 gal.	
	Sepal and leaf scab pink to petal fall	Difolatan 4F 5 qt/100 gal.	
SILVERTIP TO ½ IN. GREEN: from time the fruit buds break to show silver tip to stage when ½ in. of green tissue is protruding from buds	Scab (*Venturia inequalis*), Powdery mildew (*Podosphaera leucotricha*)	Liquid lime-sulfur 2 gal. or wettable sulfur 6 lb. If apple scab is the only problem substitute Dodine (Cyprex) 6 oz or Glyodin 1 qt. Early scab infections will frequently become established at this time on calyces of opening flower buds	
TIGHT CLUSTER SPRAY: when buds in the cluster are short-stemmed and closely packed	Scab, powdery mildew, Black Rot	Same as silvertip to ½ in. green	
FULL PINK SPRAY: when opening buds in fruiting clusters show pink	Scab, powdery mildew, cedar rust, quince rust	*For weekly spray program select one of following*	
		Captan	2 lb
		or Dodine (Cyprex)	½ lb
		or Mancozeb	2 lb
		or Maneb (Manzate D or M-22)	1½ lb
		or Thiram (Thylate)	2 lb
		or Phybam-S	4 lb
		For protective spray program (sprays applied before infection periods) select one of following:	
		Glyodin	2 pt
		or Mancozeb	2 lb
		or Niacide "A" or "M"	2 lb
		or Ferbam	2 lb
		or Wettable sulfur	4 lb
		Eradicant Sprays to be used when infections of scab are known to have occurred. Choose one of following:	
		Glyodin	1 pt
		or Dichlone (Phygon XL)	¼ lb
		or Dodine (Cyprex)	¾ lb

TABLE 6.6 *(Continued)*

<table>
<tr><th>Growth Stage</th><th>Diseases Controlled</th><th>Materials and Amounts in 100 gal. and General Comments</th></tr>
<tr><td colspan="3">From pink stage through petal fall the choice of fungicide will depend on individual grower practices and preferences. For those growers who maintain careful and accurate records of apple scab infection periods in relation to rainfall and temperature, a protective spray program may be used. For growers who do not keep apple scab spore discharge records, a 7-day (weekly) spray program should be followed. For growers who are equipped with dusting equipment, choose one of the following fungicides for apple scab protection during critical periods:</td></tr>
<tr><td></td><td></td><td>Dodine (Cyprex) 4% dust — 30 lb/acre
or Captan 7.5% dust — 40 lb/acre
or Dichlone (Phygon) 3% dust — 40 lb/acre
or Sulfur-phygon dust (Kolo 100) — 40 lb/acre</td></tr>
<tr><td colspan="3">If rust diseases are a problem use ferbam, zineb or Niacide, or add ½ lb 76% WP ferbam, or ½ lb 75% WP zineb, or 1 lb Niacide to the selected scab control chemical. NEVER APPLY CAPTAN, DICHLONE, SULFUR, OR PHYBAM-S CLOSER THAN 4-5 DAYS FROM OIL SPRAYS</td></tr>
<tr><td>BLOOM SPRAY: from the time blossoms first open until petal fall</td><td>Apple scab,
powdery mildew,
rusts,
fireblight</td><td>Same as pink spray for scab, Black rot and rusts. If fireblight is a problem make nighttime applications of Streptomycin 50-100 ppm 2-3 applications at 3-4 day intervals. Do not combine streptomycin with other fungicides.</td></tr>
<tr><td>PETAL-FALL SPRAY: when the last of the petals are falling</td><td>Apple scab,
Black Rot,
rusts,
powdery mildew</td><td>Benomyl — 4-6 oz/100 gal.
or Captan 50% WP — 1½ lb/100 gal.
or Dikar 80% WP — 1½ lb/100 gal.
or Mancozeb 80% WP — 1½ lb/100 gal.
or Polyram 80% WP — 1½ lb/100 gal.

Federal regulations do not permit further use of streptomycin</td></tr>
<tr><td>FIRST AND SECOND COVER SPRAY: 7-10 days after petal fall and 10-14 days after first cover spray</td><td>Apple scab,
Sooty Blotch and Flyspeck,
Botryosphaeria rot, etc.</td><td>Captan 50% WP — 1½ lb or
Maneb (Manzate-D, Dithane M-22) 80% WP — 1 lb or
Thiram (Thylate) 65% WP — 1½ lb
If apple scab is the only problem use Dodine (Cyprex) ⅜ lb. Dodine will not control other summer diseases. If a protective program is being followed select one of the following fungicides:
Glyodin 30% LC — 1 qt or
Ferbam 76% WP — 1½ lb or
Niacide "M" — 1½ lb</td></tr>
</table>

THIRD AND SUCCEEDING COVER SPRAYS: 8-12 days following second cover spray. Succeeding sprays should be timed according to weather conditions and disease and insect problems	Same as first cover spray	Same as first cover spray. Fungicide dosages may be reduced at this time to ½-¾ the dosage suggested for the pink spray. The number of cover sprays required for the season will be determined by weather conditions and pest control needs	

Since many of the apple insect problems develop during the summer months requiring regular applications of insecticides for their prevention, inclusion of fungicides with these applications is good insurance against summer diseases. Powdery mildew (*Podosphaera leucotricha*) may become a major problem on Rome, Cortland, Baldwin and Jonathan varieties especially when sulfur fungicides were not used early in the season. If mildew start to become established add to the other pesticides used in the cover sprays one of the following:

Dinocap (Karathane) 25% WP	2-4 oz or
Dinocap (Karathane) 48% LC	1-2 oz or
Morestan 25% WP	2 oz

Sooty Blotch and Flyspeck (*Gloeodes pomigena*) may occasionally become a late season problem, especially in low-lying sections of the orchard or in areas adjacent to a woodlot. While the fungicides suggested for apple scab control are effective for sooty blotch early in the season, they do not have the long residual action needed for sooty blotch control if the interval between the last spray and harvest exceeds 30 days. Since fruit spray residues at harvest are undesirable, special sprays for sooty blotch and flyspeck may be necessary

For Sooty Blotch control on early-maturing varieties such as McIntosh, Cortland, and Red Delicious use:

Captan 80% WP	½ lb plus Glyodin 30% LC 1 pt in 100 gal. or
Thiram (Thylate) 65% WP	1 lb plus Glyodin 30% LC 1 pt in 100 gal.

For Sooty Blotch control on late-maturing varieties such as Golden Delicious, Grimes Golden, Rome, Baldwin and Northern Spy use:

Folpet (Phaltan) 50% WP	1½ lb
or Captan 50% WP	¾ lb plus zineb 75% WP ¾ lb in 100 gal.

PRE-HARVEST SPRAY: Apple 10-14 days before harvest to fruit which will be stored after harvest	Storage Scab Storage Rots	Captan 50% WP	2 lb in 100 gal. or
		Captan 50% WP	1 lb plus 1 lb in 100 gal. Zineb 75% WP
		or benomyl 50% WP	8 oz/100 gal.

TABLE 6.7. SPECIFIC SUGGESTIONS FOR THE COMMERCIAL CONTROL OF PEAR DISEASES

Growth Stage	Disease	Control Suggestions
PRE-BLOSSOM SPRAY: When fruit buds begin to separate in the cluster	Scab (*Venturia pyrina*) Leafspot (*Fabraea maculata*)	Spray with captan 2 lb/100 gal. water or ferbam 76% WP 1½ lb/100 gal. water or benomyl 50% WP 4-6 oz/100 gal. plus recommended insecticides. Use 600-800 gal./acre for average size trees. If mildew is a problem add Karathane 25% WP ½ lb/100 gal. water
BLOOM SPRAY: from period when first blossoms open until the start of petal fall	Fireblight (*Erwinia amylovora*)	Spray with Streptomycin sulfate 100 ppm or Streptomycin sulfate 50 ppm plus Glycerin (USP grade) 2 qt/100 gal. or bordeaux mixture 2-6-100 plus recommended insecticides. The number of bloom sprays required will be determined by the severity and past prevalence of fireblight. Fireblight is favored by the occurrence of open blossoms, maximum temperatures of 65°F or higher, and a relative humidity of 60% or more. If these conditions exist, or are likely to occur within 24 hours, the use of antibiotic sprays will usually be necessary. When only light epidemics of fireblight are expected usually only one application of antibiotic will be necessary. Night applications of streptomycin will usually give better results than daytime applications. Sprays should be applied when temperatures reach 70°F. If bordeaux mixture is used it *should not* be followed by streptomycin. In severe epidemics of fireblight three sprays of antibiotic may be advisable. The first application should go on when the first blossoms open, the second when one-half of the remaining blooms are open, and a third application 3-4 days later. Streptomycin should not be applied to pears after the fruit is formed and is usually most effective when applied alone
PETAL FALL: when the last petals start to fall	Fabraea Leafspot Scab and Sooty Blotch	Spray with captan 50% WP 2 lb/100 gal. water or ferbam 76% WP 1½ lb/100 gal. water or benomyl 4-6 oz/100 gal. plus recommended insecticide use 800-1000 gal./acre for average size trees

FIRST AND SECOND COVER SPRAYS: 10-14 days after petal fall and again 10-14 days later	Same as petal fall	Same as petal fall
LATE SUMMER SPRAYS: usually applied in early August	Fabraea leafspot and sooty blotch	If leafspot and sooty blotch are still a problem at this time apply the fungicides suggested in petal fall spray
POSTHARVEST SPRAY:	Fireblight	In orchards where fireblight has been a serious problem an application of ½ lb fixed Copper 53% or ½-½-100 bordeaux mixture is advisable using 800-1000 gal./acre

TABLE 6.8. SPECIFIC SUGGESTIONS FOR THE COMMERCIAL CONTROL OF STONEFRUIT DISEASES

Growth Stage	Diseases Controlled	Materials and Amounts in 100 gal. Water and General Comments
APRICOTS		
BLOOM SPRAY: apply from balloon stage through bloom	Brown rot blossom blight (*Monilinia fructicola*)	Captan 50% WP 2 lb/100 or benomyl 50% WP 4 oz plus Captan 1-1½ lb/100 gal.
PETAL FALL: apply as soon as most of the petals have fallen	Brown rot Scab (*Cladosporium carpophilum*)	Same as bloom spray
SHUCK SPLIT: apply when shucks on developing fruit start to split	Brown Rot Scab	Same as petal fall
FIRST COVER SPRAY: apply 8-10 days after shuck split	Brown rot Scab	Same as petal fall
SECOND COVER SPRAY: 8-10 days after first cover spray	Brown rot Scab	Same as petal fall
SUMMER SPRAYS: apply as needed	Brown rot Scab	Two summer spray applications at 10-14 day intervals are generally suggested for insect control. The fungicides suggested in petal fall should be included with these insecticides. Do not apply last application closer to harvest than is permitted for the insecticide used
PLUMS AND PRUNES		
DELAYED DORMANT: apply in spring just before buds break	Black Knot (*Dibotryon morbosum*)	Spray with Zineb 65% WP 2 lb/100 when using zineb with 70-vis Superior Oil. Fill sprayer tank half-full, add zineb with agitator running, and add oil after zineb has been mixed in the sprayer tank. If oil is not used for scale and European Red Mite control, spray with liquid lime-sulfur 10 gal./100 gal.

BLOOM SPRAY: apply in early bloom	Black Knot Brown Rot (*Monilinia fructicola*)	Spray with Liquid Lime-sulfur 2 gal./100 gal. or Dichlone (Phygon) 50% WP ½ lb/100 gal. or wettable sulfur 6 lb/100 gal. or Zineb 65% WP 2 lb/100 gal.
PETAL FALL: apply when most of the petals have fallen	Brown rot Leafspot Black Knot	Spray with ferbam 76% WP 1 lb plus 3 lb wettable sulfur/100 gal. or Benlate 50% 4 oz plus Captan 50% WP 1 lb
SHUCK SPLIT: usually 10-14 days after petal fall	Leafspot, Brown Rot Black Knot	Same as petal fall
FIRST AND SECOND COVER SPRAYS: 10-14 days after shuck split and 10-14 days later	Brown rot Leafspot	Same as petal fall
THIRD AND FOURTH COVER SPRAYS: 30 days and 15 days before harvest	Brown rot Leafspot	Spray with Captan 50% WP 2 lb/100 gal. or wettable sulfur 5 lb/100 gal.
SWEET CHERRIES		
BLOOM SPRAY: apply when 10-20% of the blossoms are open	Common Brown Rot Blossom Blight (*Monilinia fructicola*)	Wettable sulfur 6 lb/100 or Dichlone (Phygon) 50% WP ½ lb/100 or Sulfur paste 6 lb/100 gal. In wet weather additional sprays or dusts of sulfur or Phygon may be necessary
PETAL FALL SPRAY: apply when most of the petals have fallen	Brown rot Leafspot (*Coccomyces hiemalis*)	Spray with Captan 50% WP 2 lb/100 gal. or ferbam 76% WP 1 lb plus 3 lb of wettable sulfur in 100 gal. If leafspot is a serious problem add dodine (Cyprex) 65% WP ¼–½ lb/100 gal. or benomyl 50% WP 4 oz plus Captan 1 lb/100 gal.
FIRST COVER SPRAY AND SECOND COVER SPRAY: 10-14 days after petal fall and 10-14 days later	Brown Rot Leafspot	Same as petal fall
THIRD COVER SPRAY: 7-10 days after	Brown rot, Leafspot	Spray with Captan 50% WP 2 lb/100 gal. or

TABLE 6.8 *(Continued)*

Growth Stage	Disease Controlled	Materials and Amounts in 100 gal. Water and General Comments
second cover or when first fruit flies emerge	Rhizopus rot	ferbam 76% WP 1 lb plus wettable sulfur 3 lb/100 gal. Where Rhizopus and Brown rot fruit rots are a problem, spray, with Captan 50% WP 1 lb plus Botran 50% WP 1 lb/100 gal.
FOURTH COVER SPRAY: apply 12-14 days later	Brown Rot Leafspot Rhizopus rot	Same as third cover spray. If brown rot is a problem at harvest captan may be applied within 24 hours of harvest. Botran may leave a yellow residue on fruit
POSTHARVEST: immediately after harvest is completed	Leafspot	Dodine (Cyprex) 65% WP ¼-⅜ lb/100 gal. or benomyl 8 oz/100 gal.
PEACHES		
DORMANT: apply in the fall after leaf drop, or in the spring before buds swell	Peach Leaf Curl (*Taphrina deformans*)	Spray with ferbam 76% WP 2 lb/100 gal. or bordeaux mixture 6-6-100 (especially on bacterial leaf spot-susceptible varieties) or *in the spring only* liquid lime-sulfur 5 gal./100
PRE-BLOOM: apply when blossoms show pink but before they open	Brown Rot Blossom Blight (*Monilinia fructicola*) Powdery mildew (*Sphaerotheca pannosa*)	Spray with Captan 50% WP 2 lb/100 gal. or liquid lime-sulfur 2 gal./100 or Dichlone (Phygon) 50% WP ½ lb/100 gal. or wettable sulfur 5 lb actual/100 gal. or Sulfur-phygon formulations (e.g., Kolo 100) at manufacturer's directions
BLOOM: apply when 5% of the blossoms are open. In wet weather reapply at full bloom	Brown rot Powdery mildew	Same as pre-bloom. If mildew is a problem use wettable sulfur alone or in combination with Captan using 2 lb sulfur plus 1 lb captan in 100 gal. water

PETAL FALL: apply when most of the petals have fallen	Brown rot Powdery mildew	Spray with Captan 50% WP 2 lb/100 gal. or Thiram (Thylate) 65% WP 2 lb/100 gal. or wettable sulfur 5 lb actual/100 gal. or sulfur paste 6 lb/100 gal.
	On varieties susceptible to bacterial spot (*Xanthomonas pruni*) a combination of dodine (*Cyprex*) 65% WP ½ lb plus 1 lb of Captan in 100 gal. may be tried starting at this time	
SHUCK SPLIT: apply 10-12 days after petal fall. May be omitted if brown rot and powdery mildew are not serious problems	Brown Rot Powdery mildew	Same as petal fall or benomyl 50% WP 4 oz plus Captan 1 lb/100 gal.
FIRST COVER: apply 10 days after shuck split or as soon as shucks have fallen	Brown Rot Scab	Same as shuck split. If peach scab is a problem wettable sulfur or sulfur pastes are the preferred materials. Dust or spray with sulfur beginning 2 weeks after bloom and continue for the next 21 days. The critical period for peach scab is shuck split stage through second cover. LIQUID LIME-SULFUR WILL NOT CONTROL PEACH SCAB. Reapply sulfur within 24 hours after each rain during this period. Be sure to cover tops and centers of trees
	Bacterial Spot (*Xanthomonas pruni*)	Varieties susceptible to bacterial spot should be sprayed with 4-6-100 zinc bordeaux (zinc sulfate 4 lb, lime 6 lb in 100 gal.)
	Powdery mildew	If mildew starts to build up continue with wettable sulfur 6 lb/100 gal. or 2 lb wettable sulfur in combination with captan or thiram
SECOND COVER: apply 10 days after first cover	Brown rot Bacterial Spot Scab	Same as first cover spray
THIRD, FOURTH AND ADDITIONAL COVER SPRAYS	Brown rot, bacterial spot	Same as first cover spray

TABLE 6.8 *(Continued)*

Growth Stage	Diseases Controlled	Materials and Amounts in 100 gal. Water and General Comments
PRE-HARVEST SPRAYS: 10-14 days after fourth cover or approximately 21 days before harvest	Brown rot and Rhizopus rot	Spray with Captan 50% WP 2 lb/100 gal. or wettable sulfur 5 lb/100 gal. or Captan 50% WP 1 lb plus Botran 75% WP 1 lb/100 gal. or benomyl 50% WP 4 oz plus Captan 1 lb/100 gal.
	Since the dates of harvest of peaches will vary greatly with different varieties special consideration should be given to the interval between final spray and harvest. These cut-off dates are as follows: Botran 1 day before harvest; Captan 0 days before harvest Wettable sulfur 0 days before harvest; Dodine (Cyprex) 15 days before harvest	

TABLE 6.9. SPECIFIC SUGGESTIONS FOR THE COMMERCIAL CONTROL OF SMALLFRUIT DISEASES

Growth Stage	Diseases Controlled	Materials and Amounts in 100 gal. Water and General Comments
GRAPES		
FIRST COVER SPRAY: apply when new shoots are 2-4 in. long	Black Rot	Captan 50% WP 2 lb/100 gal. or ferbam (Fermate) 76% WP 1½ lb/100 gal.
SECOND AND THIRD COVER SPRAY: apply when new shoots are 8 in. long and just before bloom	Black Rot	Same as first cover spray
FOURTH COVER SPRAY: apply just after bloom	Black Rot Downy Mildew (*Plasmopara viticola*)	Benomyl 50% WP 8 oz/100 gal. or Captan 50% WP 2 lb/100 gal. or Ferbam 76% WP 2 lb/100 gal. To aid in the control of black rot and downy mildew, control weeds under vines. If powdery mildew appears switch to copper fungicides suggested in fifth cover spray
FIFTH COVER SPRAY: apply in mid-season when fruits are the size of marbles	Black Rot Downy mildew Powdery mildew (*Uncinula necator*)	Spray with Fixed Copper (e.g., Tribasic, C.O.C.S.) at manufacturer's directions. Repeat if black rot or powdery mildew begins to spread
RASPBERRIES		
DELAYED DORMANT: apply when tips of buds show green	Anthracnose (*Elsinoe veneta*)	Liquid lime-sulfur 10 gal./100 plus a suitable wetting agent. On Blackcaps or in plantings where anthracnose has been especially severe, reapply after 4 days and before new leaves are 1 in. long
PRE-BLOOM: apply 7-10 days before bloom	Anthracnose Spur Blight (*Didymella applanata*) Cane Blight (*Leptosphaeria coniothyrium*)	Captan 50% WP 2 lb/100 or Folpet (Phaltan) 50% WP 2 lb/100
POST BLOOM: apply immediately after bloom	Anthracnose	Same as pre-bloom. For adequate anthracnose control old fruiting canes, or badly diseased

TABLE 6.9 *(Continued)*

Growth Stage	Diseases Controlled	Materials and Amounts in 100 gal. Water and General Comments
		canes should be pruned out and burned immediately after harvest. To prevent spread of virus diseases, separate red and black raspberry plantings by at least 300 ft, and control insects such as aphids and mites with recommended insecticides
STRAWBERRIES		
PRE-BLOOM: apply when the first blossoms start to open	Leafspot (*Mycosphaerella fragariae*) Leaf blight (*Dendrophoma obscurans*) Fruit Rots (*Rhizopus* sp., *Botrytis* sp.)	Thiram (Thylate) 65% WP 2 lb/100 gal. or Captan 50% WP 2 lb/100 gal. Dust formulations of the above are available and may be substituted for sprays at the rate of 30-40 lb/acre. In wet springs when blossoms open slowly an additional application of the above materials 7 days after the first application may be necessary or benomyl 50% WP 8 oz/100 gal.
EARLY BLOOM: apply when 10% of the blossoms are open	Leafspot, Leaf blight Fruit rots	Same as pre-bloom
PETAL FALL: apply when most of the petals have fallen	Leafspot, Leaf blight Fruit rots	Same as pre-bloom
PRE-HARVEST: apply 10 days before first picking	Leafspot, Leaf blight Fruit rots	Same as pre-bloom
DURING HARVEST PERIOD	During harvest rainy periods will favor Gray Mold (*Botrytis*) fruit rot development. If Thiram is used within 3 days of picking residues must be removed by washing. Captan may be applied up to 24 hours before harvest or benomyl 50% WP 4 oz/100 gal.	
LATE FALL SPRAY: apply just before fall application of mulch	Leaf spot Leaf Blight	Benomyl 50% WP 8 oz/100 gal. or Captan 50% WP 2 lb/100 gal.

TABLE 6.10. HARVEST RESTRICTIONS FOR FRUIT FUNGICIDES

Fungicide	Day Between Final Application and Harvest								
	Apples	Blueberries	Cherries	Grapes	Plums and Prunes	Pears	Peaches	Raspberries	Strawberries
Benomyl (Benlate)	0	21	0	7	0	0	0	3	0
Bordeaux	0	0	0	0	0	0	0	0	0
dicloran (Botran)	—	—	1	1	*	—	1	—	—
Captan	0	0	0	0	0	0	1**	0	0
captafol (Difolatan)	**	21	**	—	**	—	**	—	—
dichlone (Phygon)	1	—	3	—	3	—	7	—	3
Dikar	21	—	—	—	—	—	—	—	—
dinocap (Karathane)	21	—	—	21	—	21	45	—	—
dodine (Cyprex)	7**	—	0**	—	—	7**	15**	—	14
ferbam (Fermate)	7	40	0	7	7	7	21	40	14
folpet (Phaltan)	0	0	0**	0	—	—	—	0	0
Glyodin	0	—	7	—	—	0	0	—	—
Lime-sulfur	0**	—	0**	0**	0**	0**	0**	0**	—
mancozeb (Dithane M-45)	21–30**	—	—	66**	—	15**	—	—	—
Niacide M	0	—	—	—	—	—	—	—	—
Polyram	15	—	—	—	—	—	—	—	—
Sulfur	0	0	0	0	0	0	0	0	0
thiram (Thylate)	0	—	—	—	—	—	7	—	3**
Streptomycin	50	—	—	—	—	30	—	—	—
zineb	15	—	7	7	30	7	30	14	7

* = use in pre-bloom or postharvest only.
— = not approved for use.
** = apply in dormant only or refer to label restrictions.

One of the first reports of acquired resistance to an organic fungicide originated from the apple orchards of New York state where dodine (Cyprex) had been used exclusively for apple scab control. In apple orchards where this fungicide had been used for approximately ten years, it was noticed that it suddenly failed to control the apple scab disease. In laboratory experiments with dodine-tolerant strains of the apple scab fungus (*Venturia inequalis*) it was discovered that they would tolerate 3–4 times the dosage of dodine that previously would inactivate the apple scab fungus. Acquired resistance to dodine was confirmed, and commercial fruit growers of New York state were forced to change to other fungicides for the control of this apple disease.

When benomyl (Benlate) was first introduced as a fungicide (1968), it was highly fungicidal against most fruit pathogens. However, in 1975 benomyl-resistant strains of the apple scab, peach brown rot, and cherry leafspot pathogens were reported from several commercial orchards in Michigan. In the fruit-producing region of São Paulo, Brazil, reports were circulating as early as 1974 that benlate was not as effective for the control of strawberry leafspot disease as it was originally.

Where resistance to benomyl or dodine has been observed, or where continued use of these fungicides no longer gives satisfactory disease control, two measures may be suggested for preventing build-up of resistant strains of fruit pathogens. If benomyl or dodine has been used continuously for several seasons, it is suggested that it be substituted with non-related fungicides. The second alternative is to use benomyl in tank-mix combination with non-related fungicides such as suggested below.

Benomyl 50%	WP	2–3 oz/100 gal.
plus Captan 50%	WP	12–16 oz/100 gal.

or

Benomyl 50%	WP	2–3 oz/100 gal.
plus Mancozeb		12 oz/100 gal.
or plus Polyram 80%	WP	16 oz/100 gal.

Evidently, the development of strains of fruit pathogens tolerant or resistant to some fungicides is a problem of increasing importance. To cope with the possibility of emergence of fungi resistant to organic fungicides, the commercial fruit grower should be aware of the problem and should be prepared to adjust his selection of fungicides to avoid the build-up of fungicide-resistant strains of fruit pathogens.

FLOWER DISEASE CONTROL

Most flowers are subject to attack by a variety of disease-inciting agents which may mar the beauty of colorful blossoms, result in loss of leaves, and in some cases completely destroy annual and perennial flowering plants. The tulip, the crocus, and the jonquil are always welcome harbingers of spring after the long winter months and, together with the peony, provide the first colorful addition to the average home flower garden. For the amateur gardener who has long anticipated this awakening of plant life in the home garden, the early appearance of tulip fire, peony blight, and other flower diseases is a reminder that most garden plants may fall victim to destructive plant diseases if plans for their control are omitted from the annual gardening scheme.

Commercial growers of roses, carnations, chrysanthemums, and other sources of cut flowers are frequently confronted with numerous destructive plant diseases. Most commercial flowers are produced in greenhouses where conditions frequently favor the development and rapid spread of plant diseases which are difficult to eliminate once they have become well established.

By far the greatest number of flower diseases are caused by pathogenic molds, or fungi, which may cause root rots, stem rots, blossom blights and leafspot diseases. Many fungi can survive in the soil on infected plant tissue. For example, powdery mildew, which is a common problem in the home garden and in commercial greenhouses on roses, chrysanthemums, and other plants, can survive only from season to season on previously infected plant tissue. Rust diseases of carnations, hollyhocks, chrysanthemums, and lilies are examples of diseases caused by rust fungi.

Bacteria, commonly referred to as "microbes" by some laymen, are responsible for flower diseases such as crown gall, bacterial scab of gladiolas, leafspot of geraniums, and soft rot of hyacinth bulbs. The bacterial diseases of flowers, like bacterial diseases of humans, are highly contagious and generally difficult to control and eliminate once they become established.

Before the invention and perfection of the electron microscope, virus diseases of humans, animals, and plants were known to exist but their nature was imperfectly understood. Now, with the aid of the electron microscope in many departments of plant pathology, viruses causing plant diseases can be identified by observing the virus particle under the electron microscope. Viruses are now recognized and understood as the causal agents of flower diseases such as stunt disease of chrysanthemums, rose mosaic, lily mottle, and tulip breaking.

The general principles of plant disease control previously described are just as important for the control of flower diseases as they are for fruit and vegetable disease control. A knowledge of these plant disease control principles is essential for successful and economical prevention of flower diseases. Most commercial florists practice an integrated plant disease control program, but some amateur flower growers neglect to use these principles and expect to control plant disease problems after the problem appears by the use of miracle fungicides to cure the problem. Unfortunately, there are no miracle chemical cures or preventatives for the control of flower diseases. However, there are many safe and effective organic fungicides now available for the control of flower diseases, provided the user is familiar with their virtues and shortcomings and is willing to use them as supplementary tools for plant disease prevention.

Before considering the use of fungicides for control of flower diseases the reader should familiarize himself with the accepted Golden Rules of Flower Disease Control given below:

1. Use ONLY disease-free seed or planting stock.
2. Grow ONLY locally adapted plant types and varieties and avoid exotic types advertised in some catalogs.
3. Follow regular sanitation practices such as cleaning up and removing diseased plant debris, blighted blossoms, and other infected refuse.
4. Learn to accurately diagnose the disease problem or consult extension specialists for correct disease diagnosis.
5. Always include a plant disease control plan in the flower production program, and be prepared to follow suggested control practices as soon as the disease appears or before it becomes widely established.
6. If chemical fungicides become necessary for the control of flower diseases, apply products registered for use on the plant types concerned, at the manufacturer's recommended dosage, and using efficient application equipment. Reapply the fungicides as needed, observing recommended safety precautions.

General Considerations Concerning Chemical Control of Flower Diseases

The use of chemicals (fungicides) for disease control of ornamentals is only *ONE* of the methods of plant disease control to be used in an integrated program of plant protection. There are no miracle chemical controls for any of the common diseases of ornamentals. Use of fungicides on ornamentals by the professional grower and the amateur gardener is governed by the provisions of the Federal Insecticide, Fungicide and Rodenticide Act (FIFRA 1948), and by the Federal Environmental Pesticide Control Act (FEPCA) and its amendments approved by Congress in 1972. Under the provisions of FEPCA it has been declared that it is a federal offense "if any pesticide is used by a grower in any manner inconsistent with the label of the specific package of the pesticide being used. Where more than one company packages the same chemical the labels may vary considerably. Different formulations have different labels and these will change at irregular intervals." The Federal Insecticide, Fungicide and Rodenticide Act, as amended, states that separate registrations are necessary for each use of a pesticide. Therefore, unless greenhouse crops are mentioned on a label, the material is not registered for that specific use.

Before using any of the fungicides suggested in the following summaries, the commercial florist and the amateur flower gardener should consider carefully the following suggestions:

1. Identify the cause of the disease problem, or have it diagnosed accurately by the plant disease clinic of your local land-grant university, by extension personnel in your region, or by competent professional plant pathologists in your state experimental station.
2. Select the specific fungicide listed as effective against the pest or pathogen concerned and make sure it is registered for use on the crop, plant, or commodity involved.
3. Check the mammalian toxicity of the fungicide selected, and if possible avoid highly toxic materials. Practice the precautions specified on the product label. (Refer to Chapter 11, The Toxicity and Safe Use of Fungicides, for specific information on mammalian toxicity.)
4. It has been estimated that the average cost of developing a fungicide is in the vicinity of $1.5–3.0 million. Much of this investment is tied up in the cost of mammalian toxicity tests required for registration of the product and the preparation of an approved label for the container. The label on any pesticide product is an expensive item and contains a lot of useful information. For this reason it is recommended that all users of these products READ THE LABEL AND PRACTICE THE PRECAUTIONS SUGGESTED.
5. If in doubt, DON'T GAMBLE WITH FUNGICIDES. Obtain competent professional advise BEFORE TAKING CHANCES.
6. In view of the foregoing comments, specific dosage recommendations for fungicides for the control of diseases of ornamentals are not prescribed in this publication. Varying conditions in different geographical areas, and constantly changing federal pesticide regulations make it impossible to suggest universal recommendations for flower disease control. Instead, the effective fungicides for the control of common flower diseases are listed in the following tables. This listing is based on the manufacturer's label and on the opinions of qualified land-grant university specialists. Also summarized is a listing of the registered uses of the product and the common name, trade names, and specific disease control uses of the product. Recom-

mended dosages of the various fungicides are given on the package or container.

After selecting the fungicide for the disease problem involved, the reader should read the label on the package; if additional information is needed he should contact the extension plant pathologist at the land-grant university or the local county agricultural extension agent.

Effective Fungicides for the Prevention of Flower Diseases

Bacticin is an antibiotic suggested as a preventative and curative treatment for bacterial tumors on woody plants. It is available as a paste emulsion containing 0.93% active ingredient (a mixture of 2,4-xylenol and meta cresol).

Banrot is a relatively new product consisting of a mixture of 15% terrazole and 25% Topsin M. It is suggested for the control of root rots caused by Fusarium, Pythium, Phytophthora, Rhizoctonia, and Thielaviopsis.

Benomyl is a systemic fungicide effective against powdery mildews and Botrytis blight. It has been used successfully for the control of Rhizoctonia root rot and damping-off. It may also be used as a soil drench for the control of crown rots, root rots, and stem rots caused by Botrytis, Cylindrosporium, Fusarium, Sclerotinia, and Thielaviopsis. It is available as a 50% wettable powder.

Captan is a broad spectrum fungicide restricted to pre-plant soil treatment in ornamental greenhouses for the prevention of damping-off and root rots. It may also be used as a foliar fungicide for specific ornamentals. Available as 50% and 80% wettable powders, 4 lb/gal. flowable formulation, and 3.5–75% dusts, it is also distributed under trade names such as Orthocide, Merpan, and Vondcaptan.

Chlorothalonil is effective for the control of Botrytis blight and certain leafspot problems. It is available as a 70% wettable powder, a 6 lb/gal. flowable formulation, 90% tablets, and a 20% exotherm powder. It is distributed under trade names such as Daconil 2787 and Exotherm Termil. The latter is an exothermic powder used as a smoke bomb for space fumigation.

Cycloheximide is an antibiotic distributed under trade names such as Acti-Dione, Acti-Spray, and Naramycin. It is effective against certain rusts and powdery mildews and is formulated as a 0.027–2.26% wettable powder.

Dinocap is a specific for powdery mildews, and is also a suppressant for spider mites. Distributed under trade names such as Karathane, Iscothane, and Capryl, it is formulated as a 25% wettable powder and a 4 lb/gal. emulsifiable concentrate.

Ethazol is a relatively new product distributed under trade names such as Aaterra, Koban, Terrazole, and Truban, and is suggested as a soil drench for the control of water molds (*Pythium* and *Phytophthora*). It is available as 30% and 35% wettable powders, 0.5% and 25% emulsifiable concentrates, and 2.5% dust.

Fenaminosulf is a soil drench fungicide distributed under the trade name Dexon for water mold control. It is available as 35% and 70% wettable powders, 5% granules, and 2.5–10% dusts.

Ferbam is one of the first carbamate fungicides and is registered for the prevention of numerous leafspots, rusts, and blights. Distributed under trade names such as Fermate, Carbamate, Ferberk, Hokmate, and Niacide, it is available as 3–98½% wettable powders and 6–25½% dusts. Use of this product is limited by an objectionable black residue that may persist on leaves and blossoms.

Fixed Coppers are sometimes referred to as "low-soluble coppers," or "bordeaux substitutes." Due to their tendency to cause injury, they have been largely replaced by organic fungicides.

Folpet is an analog of Captan and is effective for the control of powdery mildews and leafspot diseases. It is distributed under trade names such as Folpan, Phaltan, and Thiophal as 50% and 75% wettable powders and 5% and 10% dusts.

Mancozeb is a coordination product of maneb and zinc ion, and is distributed under trade names such as Manzate 200, Mancofol, Dithane M-45, and Fore Etc. These products are effective for the control of some leaf blight and rust diseases of ornamentals, and are available as 80% wettable powder and as 4.8–15% dusts.

Maneb is one of the earlier dithiocarbamate fungicides and is suggested for the control of blights and leafspots with a wider range of usefulness than Mancozeb. Distributed under trade names such as Dithane M-22, Tersan L.S.R., Polyram M, Manzate, Manoc, and Tubothane, it is formulated as 70% and 80% wettable powders, 1–20% dusts, and as a 4 lb/gal. emulsifiable concentrate.

Oxycarboxin is a systemic fungicide available as a 5% liquid formulation and as a 75% wettable powder. It has been a very effective control for carnation rust, and may be applied through injector systems. It is distributed under the trade names Plantvax and DCMOD.

Parinol is a pyridine compound which is a specific protectant against powdery mildew. It will not eradicate well established mildew infections. Distributed under the trade name Parnon, it is formulated as a 4% emulsifiable concentrate.

PCNB is also known by the common name Quintozene, and is distributed under a variety of trade names such as Terraclor, Brassicol, Botrilex, Fungiclor, and Kobutol. It is suggested for use as a pre-plant soil drench for the control of some stem rot and root rot fungi. It is not effective against root rots caused by Pythium or Phytophthora. It is available in the form of 10% granules, 0.75–1.63% pastes, 35–75% wettable powders, and 1.4–99% dusts.

Piperalin is both a protective and eradicant fungicide for powdery mildew and is sold under the trade name Pipron as an 8 lb/gal. emulsifiable concentrate.

Streptomycin is an antibiotic suggested for the control of diseases of bacterial etiology such as bacterial blight of chrysanthemums, bacterial stem rot of carnations and gardenias, and bacterial stem rot of geraniums. Sold under trade names such as Agrimycin, Hokko-mycin, Phytomycin Agri-Strep, and Fimocidin, it is available as 8.5% and 62.5% wettable powders and 0.1–0.2% dusts.

Sulfur is still used for the suppression of powdery mildews. It may be obtained as 95% dust, 30–90% wettable powders, or as 50% colloidal paste.

Zineb is a carbamate fungicide suggested for the control of blights, leafspots, and rust diseases. Application as a foliar spray with reapplication at 7–10 day intervals is recommended. Distributed under trade names such as Parzate, Dithane Z-78, Zimate, Polyram Z, and Zinosan, it is sold as a 4 lb/gal. flowable formulation, 1.4–75% wettable powder, and as 3.25–15% dusts.

ORNAMENTAL SHRUB, VINE, AND SHADETREE DISEASES

Ornamental shadetrees, shrubs, and ground cover plants are important components of every home landscape, city park, and industrial landscape plan. However, the mighty oak, the attractive hard maple, the ornamental barberry, and the ground cover myrtle frequently have their plant disease problems. The loss of valuable shadetrees, shrubs, and vines is frequently a major disappointment in many home gardens and city park developments. Plates 6 and 7 illustrate some common ornamental and shade and forest tree diseases.

Most of the common diseases of trees, shrubs and ground cover plants may be prevented, but they are virtually impossible to eliminate once they are well established. There is no panacea for all the ills of these valued plant inhabitants, but a combination of good horticultural practices and judicious use of protective fungicides will prevent these problems if implemented before the establishment of plant diseases.

In a recent survey of diseases of ornamental shrubs C.C. Powell, Jr. (1976) of Ohio State University, stated that of the 296 diseases listed by Pirone (1970) there were 681 items on the control column referring to the use of fungicides. Fungicides have been accepted by most plant pathologists as necessary preventatives for diseases of shadetrees, shrubs, and vines to be used in combination with recommended methods of efficient maintenance.

TABLE 6.11. REGISTERED USES OF FUNGICIDES FOR FLOWERS (AS OF MARCH 1, 1978)

Fungicide	Common Name	Other Trade Names	Registered for Use on
Aaterra		*See* Terrazole	
Acti-Dione	Cycloheximide	Actispray, Hizarocin, Kaken, Naramycin	Begonia, Chrysanthemum, Honeysuckle, Magnolia, Phlox, Roses, Zinnia
Agrimycin	Streptomycin	Agri-Strep, Hokko-Mycin, Phytomycin, Rimocidin, etc.	Chrysanthemum, Gladiolus, Rose
Agri-Strep		*See* Streptomycin	
Banrot	none	none	Bedding plants, Chrysanthemum, Geranium, Lily, Marigold, Petunia, Poinsettia, Rhododendron, Zinnia
Benlate	Benomyl	Grex, Tersan 1991, Ultra-Sofril	Ageratum, Aster, Azalea, Begonia, Belladonna, Caladium, Calceolaria Calendula, Calla, Camellia, Carnation, Chrysanthemum, Cineraria, Coleus, Columbine, Cosmos, Cyclamen, Daffodil, Dahlia, Daisy, Delphinium, Foxglove, Fuschia, Gardenia, Geranium, Gladiolus, Hollyhock, Hyacinth, Hydrangea, Iris, Lilac, Lily, Marigold, Morning Glory, Narcissus, Orchids, Pansy, Phlox, Poinsettia, Poppy, Primrose, Rhododendron, Rose, Stock, Snapdragon, Sunflower, Sweet Pea, Tulip, Violet, Zinnia
Benomyl		*See* Benlate	
Brassicol		*See* Terraclor	
Capryl		*See* Karathane	
Captan	Captan	Captane, Merpan, Orthocide, Vondcaptan	Ageratum, Ajuga, Alyssum, Aster, Azalea, Begonia, Caladium, Calceolaria, Calendula, Calla, Camellia, Carnation, Cineraria, Coleus, Crocus, Cyclamen, Daffodil, Dahlia, Daisy, Delphinium, Foxglove, Fuschia, Gardenia, Gladiolus, Hibiscus, Hydrangea, Hyacinth, Lily, Marigold, Morning Glory, Narcissus, Orchid, Pansy, Peony, Petunia, Phlox, Poinsettia, Poppy, Primrose, Rose, Snapdragon, Stocks, Sunflower, Violet, Zinnia
Carbamate		*See* Fermate	
Chlorothalonil		*See* Daconil	
Crotothane		*See* Karathane	
Cycloheximide		*See* Acti-Dione	

TABLE 6.11 *(Continued)*

Fungicide	Common Name	Other Trade Names	Registered for Use on
Daconil	Chlorothalonil	Bravo W-75, Exotherm, Termil	Ageratum, Aster, Begonia, Caladium, Chrysanthemum, Coleus, Delphinium, Geranium, Gloxinia, Iris, Lobelia, Petunia, Primrose, Rose, Zinnia
Dexon	Fenaminosulf	Dapa	Ageratum, Aster, Azalea, Begonia, Caladium, Calendula, Camellia, Carnation, Chrysanthemum, Coleus, Cosmos, Cyclamen, Daffodil, Dahlia, Daisy, Delphinium, Foxglove, Gardenia, Geranium, Gladiolus, Hollyhock, Hyacinth, Hydrangea, Iris, Lily, Lobelia, Marigold, Morning Glory, Narcissus, Nasturtium, Pansy, Peony, Petunia, Phlox, Poppy, Primrose, Rose, Salvia, Snapdragon, Sunflower, Sweet Pea, Tulip, Violet, Zinnia
Dinocap		*See* Karathane	
Dithane M-22		*See* Manzate	
Dithane M-45	Mancozeb	Fore, Mancofol, Manzate 200	Azalea, Camellia, Carnation, Chrysanthemum, Dahlia, Gladiolus, Holly, Hollyhock, Honeysuckle, Iris, Lily, Peony, Rose, Snapdragon, Tulip, Zinnia
Dithane Z-78	Zineb	Parzate, Polyram-Z, Sperlox-Z, Zinosan, Zebtox, Hexathane	Aster, Azalea, Camellia, Carnation, Chrysanthemum, Cyclamen, Dahlia, Delphinium, Gladiolus, Hydrangea, Iris, Lily, Orchid, Pansy, Peony, Rose, Snapdragon, Tulip, Zinnia
Ethazol		*See* Terrazole	
Exotherm		*See* Daconil	
Fenaminosulf		*See* Dexon	
Ferbam		*See* Fermate	
Ferberk		*See* Fermate	
Fermate	Ferbam	Carbamate, Ferberk, Fermocide, Hokmate, Niacide, Trifungol	Aster, Azalea, Begonia, Camellia, Carnation, Chrysanthemum, Columbine, Cyclamen, Dahlia, Delphinium, Gardenia, Geranium, Gladiolus, Hollyhock, Hydrangea, Iris, Marigold, Peony, Phlox, Poinsettia, Poppy, Primrose, Rose, Snapdragon, Sweet Pea, Zinnia
Folpan		*See* Phaltan	
Folpet		*See* Phaltan	

Fore		*See* Dithane M-45	
Fungiclor		*See* Terraclor	
Iscothan		*See* Karathane	
Karathane	Dinocap	Caprane, Capryl, Crotothane, DNOCP, Iscothan, Mildex	Aster, Begonia, Calendula, Chrysanthemum, Dahlia, Delphinium, Hydrangea, Larkspur, Snapdragon, Zinnia
Koban		*See* Terrazole	
Mancofol		*See* Dithane M-45	
Mancozeb		*See* Dithane M-45	
Maneb		*See* Manzate	
Manzate	Maneb	Dithane M-22, Tersan-LSR, Polyram M, Remasan, Tubothane	Aster, Azalea, Begonia, Camellia, Carnation, Chrysanthemum, Dahlia, Geranium, Gladiolus, Hydrangea, Iris, Lily, Pansy, Peony, Rose, Snapdragon, Stock, Zinnia
Mildex		*See* Karathane	
Niacide		*See* Fermate	
Orthocide		*See* Captan	
Oxycarboxin		*See* Plantvax	
Parinol		*See* Parnon	
Parnon	Parinol	None	Rose, Zinnia
Parzate		*See* Dithane Z-78	
PCNB		*See* Terraclor	
Phaltan	Folpet	Folpan, Thiophal	Aster, Azalea, Carnation, Chrysanthemum, Gladiolus, Iris, Marigold, Phlox, Poinsettia, Rose, Snapdragon, Zinnia
Pipron	Piperalin	None	Chrysanthemum, Dahlia, Rose, Zinnia
Plantvax	Oxycarboxin	DCMOD	Carnation
Polyram M		*See* Manzate	
Quintozene		*See* Terraclor	
Rimocidin		*See* Agrimycin	
Streptomycin		*See* Agrimycin	
Terraclor	PCNB and Quintozene	Brassicol, Fungiclor, Kobutol, Tri-PCNB, Tritisan	Aster, Azalea, Camellia, Chrysanthemum, Delphinium, Gladiolus, Hyacinth, Lily, Poinsettia, Rose, Snapdragon, Sweet Pea, Tulip, Violet
Terrazole	Ethazol	Aaterra, ETCMTB, Etridiazol, Koban, Truban	Aster, Begonia, Calendula, Carnation, Celosia, Chrysanthemum, Coleus, Dahlia, Foxglove, Geranium, Lily, Marigold, Pansy, Petunia, Poinsettia, Rhododendron, Snapdragon, Zinnia

TABLE 6.11 *(Continued)*

Fungicide	Common Name	Other Trade Names	Registered for Use on
Tersan, LSR		*See* Manzate	
Tritisan		*See* Terraclor	
Truban		*See* Terrazole	
Zineb		*See* Dithane Z-78	

TABLE 6.12. SPECIFIC USES OF FUNGICIDES FOR FLOWERS

The products summarized below are registered for the uses specified as of March 1, 1978

Disease	Cause	Effective Fungicides[1]	Comments
Damping-off	Pythium sp. Phytophthora sp.	Benomyl (Benlate)	Suggested for Rhizoctonia and Botrytis rots as a soil drench
Cutting rots	Rhizoctonia sp. Botrytis sp.	Captan	Registered for pre-plant use for damping-off
		Banrot	Suggested for damping-off
		Dexon	For Pythium and Phytophthora control
		Ethazol (Truban, Terrazole)	For Pythium and Phytophthora control
		PCNB (Terraclor)	Suggested for Rhizoctonia only
Root rots (water molds)	Pythium sp. Phytophthora sp.	Captan	Pre-plant soil drench only
		Banrot Dexon Ethazol (Truban, Terrazole)	Use as a soil drench in a pre-plant treatment program
Other root and stem rots	Cylindrosporium sp. Fusarium sp. Rhizoctonia sp. Thielaviopsis sp.	Benomyl (Benlate)	Especially effective for Botrytis
		Banrot PCNB (Terraclor)	Use as a pre-plant treatment
Leafspots and Flower Spots	Botrytis sp.	Benomyl (Benlate) Chlorothalonil (Daconil)	Especially effective for Botrytis Blight
	Septoria sp.	Captan Ferbam Mancozeb Maneb Folpet Zineb Ziram Fixed Coppers	For Septoria and Alternaria leafspots
Rusts	Uromyces (Carnation)	Oxycarboxin (Plantvax)	For Uromyces (Carnation) rust only

TABLE 6.12 *(Continued)*

The products summarized below are registered for the uses specified as of March 1, 1978

Disease	Cause	Effective Fungicides[1]	Comments
	Phragmidium (Rose) Puccinia (Chrysanthemum, Hollyhock, Iris, Snapdragon)	Ferbam Mancozeb Zineb	For Puccinia and Phragmidium rusts. Repeated applications are usually necessary
Powdery mildews	Oidium (African Violet, Snapdragon)	Benomyl (Benlate) Cycloheximide (Acti-Dione PM)	Repeated applications at 7-10
	Erysiphe (Chrysanthemum, Cosmos, Dahlia, Hydrangea, Phlox, Sweet Pea) Microsphaera (Lilac) Sphaerotheca (Rose)	Dinocap (Karathane) Parinol (Parnon) Piperalin (Pipron) Sulfur	day intervals are necessary. Addition of a sticker-spreader will improve control
Vascular Wilts	Fusarium sp. (Aster, Carnation, Chrysanthemum, Dahlia, Delphinium, Stock, Sweet William) Verticillium sp. (Chrysanthemum, Dahlia, Lilac, Peony, Rose, Snap-dragon, Stock, Strawflower)	There are no effective chemical controls for vascular wilts	
Bacterial Diseases	Erwinia sp. Bacterial blight of chrysanthemums Pseudomonas sp. (Carna-tion, Delphinium, Gardenia, Lilac) Xanthomonas sp. (Begonia, Dieffenbachia, English Ivy, Gardenia, Geranium, Poppy, Iris, Stock)	Streptomycin (Agri-Strep) Fixed Coppers	
Virus Diseases		There are no effective chemical controls for virus diseases	

Source: adapted from Powell and Lundquist (1976).
[1]For suggested dosage rates refer to the manufacturer's recommendation on the package or container.

The basic horticultural practices that must be used in partnership with fungicides for the prevention of diseases of shrubs and vines are as follows:

1. Plant only disease-free, disease-resistant, locally-adapted varieties.
2. Plant winter hardy plant types in well drained, sunny locations.
3. Supply necessary fertilizer and water requirements according to the suggestions of local qualified specialists and professional nurserymen.
4. Practice regular sanitation measures such as collection and burning of diseased leaves, twigs, and branches, and prune out cankered and diseased wood in the spring.
5. Protect pruning wounds with a recommended wound dressing.
6. Use only fungicides approved by state and federal agencies for use on shadetrees, shrubs, and vines.

The same precautions suggested for the use of fungicides on other crops must be followed in the application of fungicides for the control of diseases of shadetrees, shrubs and vines. Whether fungicides are applied as sprays or dusts they must be applied *before* the initiation of plant diseases, protecting all susceptible plant parts. It is also important to reapply protective fungicides at regular intervals as determined by weather conditions. There is no one fungicide that will effectively prevent all of the diseases of trees, shrubs, and vines. In most cases specific fungicides are suggested for individual plant disease problems.

Fungicides Approved for Use on Shadetrees, Shrubs, and Vines

ACTI-DIONE is an antibiotic distributed as a 2.6% wettable powder and suggested for the control of powdery mildew, cherry leafspot, Hawthorn leaf blight, and cedar apple rust.

BENOMYL is the common name assigned to the systemic fungicide distributed under the trade name *Benlate.* Benlate is a 50% wettable powder suggested for the control of anthracnose, leafspot disease, and powdery mildew.

Bordeaux mixture is a combination of Copper sulfate (Blue Vitriol) and fresh hydrated lime in water. Formulations are expressed as 4-4-50, 8-8-100, 10-10-100, etc. 4-4-50 means 4 lb copper sulfate and 4 lb hydrated spray lime in 50 gal. of water and is prepared as follows:

For each pound of copper sulfate in the 100 gal. formula (4-4-50 is the same as 8-8-100), use ⅓ tbsp in each gallon of water, and for each pound of hydrated lime use 1 tbsp/gal. water. For example: to make 1 gal. of 8-8-100 bordeaux mixture, first dissolve 2⅔ tbsp copper sulfate in ½ gal. water. Then dissolve 8 tbsp hydrated lime in ½ gal. water. Next, mix together the solutions of copper sulfate and lime to form the gelatinous protective coating known as "bordeaux mixture." The mixture should be freshly prepared and old mixtures should be discarded.

For the homeowner who does not wish to bother with the preparation of "home-made bordeaux mixture," there are widely available commercial preparations known as "ready prepared bordeaux mixture."

Captan is a broad spectrum fungicide distributed under trade names such as Orthocide Garden Fungicide, Captan 50W, and Orthocide 50W, and is suggested for a wide range of leafspot diseases.

Ferbam is a black wettable powder containing 76% and 96% active ingredient and distributed under trade names such as Fermate, Ferberk, and Coromate. It was one of the earliest carbamate fungicides, and is suggested for the control of cedar apple rust, grape black rot, peach leaf curl, and raspberry anthracnose.

Fixed copper is the name given to fungicides containing copper salts with 25–55% metallic copper. They are sold as wettable powders and as dust formulations. They are distributed under names such as Tri-Basic copper sulfate, C.O.C.S., Copper Compound "A," and Basicop etc., and are recommended for the control of diseases such as ash anthracnose, barberry bacterial spot, needle and twig blight of fir, and Cercospora needle blight of red cedar.

Karathane is a specific fungicide for the control of powdery mildew and is ineffective against other shadetree and shrub diseases. It is available as a 25% wettable powder and as an emulsifiable concentrate under names such as Mildex, Capryl, and Iscothane, and has been designated with the common or coined name Dinocap.

Lime-sulfur is a liquid fungicide sometimes referred to as Eau Grison. It is a caustic amber liquid that may cause injury to some trees and shrubs, especially under conditions of slow drying. It is suggested as a dormant or delayed dormant spray for the control of birch leaf blister, boxwood canker, and plum pockets.

Mancozeb is a zinc-ion coordinated carbamate distributed under the trade names Dithane M-45 and Manzate-200, and is suggested for the control of cedar rusts, plum black knot, crabapple scab, and fungus leafspots.

Maneb is a carbamate fungicide distributed under trade names such as Dithane M-22, Polyram-M, and Manzate as a 70% and 80% wettable powder. It is suggested for the prevention of apple, hawthorn, and quince rust, leafspot of cotoneaster, and other leafspot diseases.

Phaltan is a wettable powder sold under trade names such as Stauffer Folpet 75-W, Ortho Phaltan 50 Wettable, and Ortho Rose and Garden Fungicide.

TABLE 6.13. CHEMICAL CONTROL OF DISEASES OF TREES, SHRUBS, AND GROUND COVER PLANTS

Host	Disease	Description	Control
Ajuga	Crown rot	Brown-black discoloration of the crown. Plants turn yellow and die	In the spring drench the soil around the plants with 1 tbsp/gal. water of Terraclor 75% WP using 1 pt/sq ft soil area
Alder	Powdery mildew	White powdery mold on leaves and buds	Spray with wettable sulfur 2 tbsp/gal. water or Spray with Karathane (dinocap) 2 tsp/gal. water
Almond (flowering)	Blossom and twig blight	Sudden wilting of blossoms followed by dying of twigs	Spray with Captan 2 tbsp/gal. water. Two applications 7-10 days apart starting when blossoms first open
Arborvitae	Phomopsis needle and twig blight	Needles, twigs and smaller branches which turn brown and gradually die back from the tips	Spray at budbreak and at 10-14 day intervals with Benomyl 50% WP at manufacturer's direction
Ash	Anthracnose	Large, irregular brown spots mostly on leaf margins	Benomyl 50% WP 1½ tsp/gal. water starting at budbreak and when leaves are half-grown and again when leaves are fully developed
Aspen	*See* Poplar		
Azalea	Leafspots	Small spots on leaves. Infected leaves drop prematurely	Zineb 75% WP 1½ tbsp/gal. water or Basic Copper sulfate 2 tbsp/gal. water after flowering
	Leaf and flower gall	Leaves thickened and fleshy, turning pale green or white	Spray with zineb 75% WP 4 tbsp/gal. water or Basic Copper Sulfate 2 tbsp/gal. water just before leaves emerge
	Rust	Orange pustules on undersides of leaves	Spray with ferbam 76% WP 2½ tbsp/gal. water in July and August
	Powdery mildew	White powdery mold on leaves in late summer	Apply sulfur spray or dust when mildew appears or spray with Karathane (dinocap) 2 tsp/gal. water

TABLE 6.13 *(Continued)*

Host	Disease	Description	Control
Barberry	Anthracnose	Tan, brown or purple spots on leaves	Apply Benomyl 50% WP 1½ tsp/gal. water or Zineb 1½ tbsp/gal. water or fixed copper at manufacterer's recommendation
	Powdery mildew	White powdery mold on leaves	Spray with Benomyl 50% WP 1½ tsp/gal. water or spray or dust with sulfur or apply Karathane (dinocap) 2 tsp/gal. water when mildew appears
	Verticillium wilt	Leaves turn reddish-brown and drop prematurely: plant dies. Reddish streaks in wood of infected plants	Remove and burn infected plants and replace with resistant varieties
Basswood	*See* Linden		
Birch	Anthracnose	Small, irregular to circular brown spots on leaf margins	Same as for ash anthracnose
Bittersweet (Celastrus)	Powdery mildew	Leaves covered with a white powdery growth. Leaves turn yellow and fall	Spray or dust with sulfur at 7-10 day intervals as soon as mildew appears
Boxwood (Buxus)	Cankers and leaf spots	Small, rose-colored, waxy pustules on leaves and branches. Leaves turn yellow and become covered with small, black fungus fruiting bodies	Remove and burn infected plant parts. Spray with Bordeaux mixture or fixed copper when new growth starts in spring
Buttonbush (Cephalanthus)	Leaf spot	Small brown or black spots on leaves	Spray with ferbam 76% WP at 7-14 day intervals starting when new leaves begin to unfold
	Powdery mildew	White powdery coating on leaves in midsummer	Spray or dust with sulfur or Karathane 2 tsp/gal. of water

Catalpa	Leaf spots, spot anthracnose	Round to irregular, yellowish-brown to black spots on leaves. Dead tissue drops out, leaving holes in leaves	Gather and destroy fallen infected leaves. No effective chemical control
	Powdery mildew	White, powdery areas on leaves which drop prematurely	When mildew appears start applying Benomyl 50% WP 1½ tsp/gal. water or Karathane 2 tsp/gal. at 10-14 day intervals
	Verticillium wilt	Leaves on one or more branches turn brown and fall. Blue streaks in sapwood of infected branches	Remove severely infected branches and fertilize heavily in the spring
Cherry		*See* Stonefruit Diseases	
Chinese Elm		*See* Elm	
Chokecherry (Aronia)	Rust	Dark-colored cankers on twigs causing dieback	Prune out infected twigs. Avoid planting near southern white cedar. Spray with Captan 50% WP 2 tbsp/gal. water
Clematis	Leaf Spot	Irregular, reddish-brown spots on lower leaves which turn yellow and fall	Spray with zineb 75% WP 1½ tbsp/gal. water at 7-10 day intervals starting when new leaves first appear
	Stem rot	Brown-black lesions at soil level which girdle stem, resulting in wilting and death	Plant only disease-free plants and spray with zineb as suggested for leaf spot
Cotoneaster	Fireblight	Flowers near growing tip suddenly wilt and turn black. Twig dieback may also result	Plant only resistant varieties. Prune out infected twigs in dormant season. Sterilize pruning tools with 70% alcohol or 10% laundry bleach
Crabapple	Fireblight	New shoots appear as if scorched with fire	*See* Apple disease control
	Scab	Olive to black velvety spots on leaves and developing fruit	Follow regular spray program during growing season using:

TABLE 6.13 *(Continued)*

Host	Disease	Description	Control
			Benomyl 50% WP 1½ tsp/gal. or Captan 50% WP 2 tbsp/gal. or mancozeb 80% WP 1½ tbsp/gal.
	Powdery mildew	White powdery patches on young leaves, buds, blossoms and developing fruits	Follow spray schedule suggested for mildew control on apples
	Rust	Pale yellow-orange pustules on upper leaf surfaces	Spray in spring at early bloom time with ferbam 76% WP 2½ tbsp/gal. or thiram 65% WP 1½ tbsp/gal. or zineb 75% WP 1½ tbsp/gal.
Dogwood	Leaf spots and anthracnose	Spots on leaves with purple to dark brown borders	When new leaves first appear spray with: Captan 50% WP 1½ tbsp/gal. or benomyl 50% WP 1½ tsp/gal. Repeat at 7-10 day intervals
	Petal blight	Brown areas on fading flowers and leaves	Same as for anthracnose
	Crown canker	Canker at base of trunk near soil level. Leaves small and turn red early in fall	Prune out infected tissue in area of canker. Avoid wounding of tree trunk. Fertilize in spring
Dutchman's Pipe	Leaf spots	Circular or angular dead areas on leaves	2-3 applications of ferbam 2½ tbsp/gal. at 7-10 day intervals starting when spots first appear
Elm	Anthracnose	Irregular spots or dead areas between leaf veins	2-3 sprays at 7-10 day intervals starting when leaf buds open in spring using benomyl, zineb or ferbam at manufacturer's direction
	Dutch elm disease	Contact Cooperative Extension Service mailing room for detailed information and control suggestions	

Euonymous	Crown gall	Galls with irregular rough surfaces on roots and lower stems	Avoid planting nursery stock with galls. Dig up and burn severely infected plants
	Powdery mildew	Powdery white mold on leaves	At first sign of mildew spray with Acti-Dione PM at manufacturer's direction or benomyl 50% WP ½-1 tbsp/gal. or Karathane ½-1 tbsp/gal. or wettable sulfur 3 tbsp/gal.
Fir	Needle cast and twig blight	Brown spots on needles. Needles fade in color and drop prematurely	2-3 sprays of fixed copper at manufacturer's direction starting when new needles are half-grown
Firethorn		*See* Pyracantha	
Hackberry	Powdery mildew	Powdery white growth on both leaf surfaces	As soon as mildew appears spray with Acti-Dione PM, benomyl, or Karathane at manufacturer's direction
Hawthorn	Leaf blight	Small, angular, reddish-brown spots on leaves	As soon as new leaves unfold apply 3-4 sprays at 7-10 day intervals of: Captan 50% WP 1½ tbsp/gal. or benomyl 1 tbsp/gal. or mancozeb 80% WP 1½ tbsp/gal.
	Cedar-hawthorn rust	Pale yellow-orange spots on leaves	Apply as soon as new leaves unfold ferbam 76% WP 2½ tbsp/gal. or maneb 80% WP 1½-2 tbsp/gal.
	Fireblight	New shoots appear as if suddenly scorched by fire	When 20-25% of blossoms are open spray with Streptomycin at manufacturer's direction and repeat at 5-7 day intervals during bloom

TABLE 6.13 *(Continued)*

Host	Disease	Description	Control
	Powdery mildew	White powdery growth on leaves	When mildew appears apply: benomyl 50% WP 1½-3 tsp/gal. or wettable sulfur 2-2½ tbsp/gal. or Karathane 22.5% WP 1-2 tsp/gal.
Hickory	Anthracnose and leaf spots	Irregular, reddish-brown spots on lower leaf surface	3-4 sprays at 7-10 day intervals starting at budbreak using benomyl 50% WP 1½-3 tsp/gal. or zineb 75% WP 1½-2 tbsp/gal. or maneb 80% WP 1½-2 tbsp/gal.
Holly	Anthracnose and leaf spots	Round to irregular, yellow to brown spots with purple margins	Apply in late spring and reapply at 10-14 day intervals using: benomyl 50% WP 1½-3 tsp/gal. or mancozeb 80% WP 2 tbsp/gal.
Honeylocust	Leaf blight	Round to irregular spots on leaves and pods	As buds begin to unfold and 10 and 21 days later spray with: benomyl 50% WP 1½-3 tsp/gal. or fixed copper at manufacturer's direction or bordeaux mixture or zineb 75% WP 1½-2 tbsp/gal.
Hydrangea	Leaf spot	Brown spots on leaves near leaf margins	When new growth appears apply 3 sprays 10 days apart using benomyl 50% WP 1 tbsp/gal. or zineb 75% WP 1½-2 tbsp/gal.

	Rust	Rusty-brown pustules on leaves	Same as for leaf spots except use ferbam 76% WP
	Botrytis gray mold	Flowers covered with powdery gray masses of spores	When first symptoms appear spray with benomyl 50% WP ½-1 tbsp/gal. or botran 50% WP 1-1½ tbsp/gal.
	Powdery mildew	Leaves covered with white powdery growth in late summer	As soon as mildew appears spray with: benomyl 50% WP ½-1 tbsp/gal. or Karathane 22.5% WP ½ tbsp/gal.
Ivy (English)	Bacterial leaf spot and stem canker	Brown or black leaf spots with reddish margins. Stalk canker causes discolored leaf stalks with cankers on woody portion of vines	Remove infected plant parts and spray with bordeaux mixture or fixed copper
Juniper	Phomopsis twig and needle blight	Tips of branches turn brown and gradually die back. May be confused with winter kill. Small black fungus fruiting bodies	Remove infected branches. Thin out growth. Spray in the spring when new growth starts with benomyl 50% WP 1 tbsp/gal.
	Cedar-apple rust	Orange-colored gelatinous horns from galls in the spring	When telial horns start to protrude from galls spray with Acti-Dione at manufacturer's directions or ferbam 76% WP 2 tbsp/gal. or Polyram 80% WP 1½-2 tbsp/gal.
Kerria (Globeflower)	Leaf and twig blight (Coccomyces)	Tips of twigs turn brown and gradually die back. Small black fruiting bodies develop on infected leaves and twigs	When new leaves are ¼ in. long apply 4 sprays at 10-day intervals of ferbam 76% WP 2½ tbsp/gal.
	Phomopsis twig blight	Oval or long, irregular tan spots with raised edges on main stems	Same as for Coccomyces leaf and twig blight
Laurel (Mountain)	Mycosphaerella leaf spot	Irregular-circular grayish to white or silvery spots with purple borders	In the autumn rake up and burn infected leaves. In the spring as soon as new leaves start to develop spray with ferbam 76% WP 2½ tbsp/gal. and repeat at 7-10 day intervals until leaves are fully developed

TABLE 6.13 *(Continued)*

Host	Disease	Description	Control
	Phomopsis leaf blight	Circular, brown, zonate areas on leaves which enlarge and coalesce and may eventually involve the entire leaf	Same as Mycosphaerella leaf spot
Leucothoe (Sweetbells)	Leaf spot	Spots ½ in. in diameter generally covered with small, black fungus fruiting bodies	Rake up and burn infected leaves in the fall. Apply 3 sprays at 14-day intervals starting when new leaves are ¼ in. long using ferbam 76% WP 2½ tbsp/gal.
Lilac	Bacterial blight	In wet weather young shoots turn black on one side. Flower buds turn black and resemble frost injury	Prune out infected twigs. Thin out bushes to improve air circulation. When new growth develops apply 2-3 sprays of fixed copper at manufacturer's direction and repeat at 10-day intervals if disease persists
	Powdery mildew	Leaves and stems covered with white powdery growth	Spray at 7-10 day intervals, starting when mildew first appears, and continuing as long as mildew persists. Apply: benomyl 50% WP ½-1 tbsp/gal. or Karathane 22.5% WP ½-1 tbsp/gal. or wettable sulfur 4-6 tbsp/gal.
Linden (Basswood)	Anthracnose, leaf spots, and leaf blotch	Small to large, irregular, light brown spots with dark margins	Just after budbreak and again 10 and 21 days later spray with fixed copper at manufacturer's direction or benomyl 50% WP ½-1 tbsp/gal.
	Powdery mildew	Powdery, grayish-white patches on leaves and young shoots	When mildew first appears spray with wettable sulfur 4-6 tbsp/gal. or benomyl 50% WP ½-1 tbsp/gal.
Magnolia	Powdery mildew	Same as powdery mildew on Linden	

	Leaf blight		Two sprays 10 days apart starting at budbreak using any fixed copper at manufacturer's direction plus benomyl 50% WP ½-1 tbsp/gal.
Maple	Anthracnose	Narrow, purple to brown streaks along leaf veins of Norway maple, and large brown, irregular areas between leaf veins of sugar maple	At budbreak when leaves are half-grown spray with: benomyl 50% WP 1½ tsp/gal. or mancozeb 80% WP 1½-2 tbsp/gal. or zineb 75% WP 1½-2 tbsp/gal.
Matrimony vine (Lycium)	Powdery mildew	White powdery mold on leaves in late summer and fall	As soon as mildew appears spray or dust with wettable sulfur or spray with Karathane at manufacturer's direction
Mimosa (Silk Tree)	Fireblight	*See* Crabapple	
	Cytospora canker	Irregular cankers which frequently girdle twigs, branches or trunks of low vigor trees	Prune out infected branches. Fertilize in the spring. If the problem becomes serious spray in the spring with bordeaux mixture, or with benomyl 50% WP 1 tbsp/gal.
Mock orange	Ascochyta leaf spot	Circular brown spots on leaves	Three applications at 14-day intervals of any fixed copper fungicide at manufacturer's direction in the early spring
	Nectric canker	Cankers on stems with small, red fungus fruiting bodies on dead bark	Prune out and destroy infected twigs in dormant season
Mountain ash	Leaf spot and scab	Round spots on leaves	If serious spray 2-3 times starting as leaves begin to unfold using benomyl 50% WP ½-1 tbsp/gal. or zineb 75% WP 1½-2 tbsp/gal.
	Fireblight	Terminals and young twigs turn black and die. Appear as if scorched by fire	When 20-25% of blossoms are open spray with Streptomycin at manufacturer's direction. Repeat in full bloom

TABLE 6.13 *(Continued)*

Host	Disease	Description	Control
New Jersey tea (Ceanothus)	Powdery mildew	Leaves covered with a white mold in late summer	As soon as mildew asppears spray with wettable sulfur or benomyl
Oak	Anthracnose	Scattered brown spots along leaf veins giving leaves a scorched appearance	If severe in wet years apply bordeaux mixture, Kocide 101, or other fixed copper fungicides at early budbreak and repeat for 2 more applications at 10-day intervals
	Leaf blister (Taphrina)	Yellow or brown raised blisters on upper leaf surface	Apply 3 sprays starting just before budbreak and 2 more applications at 14-day intervals Captan 50% WP 2 tbsp/gal. or Zineb 75% WP 1½-2 tbsp/gal. or maneb 80% WP 1½-2 tbsp/gal.
	Wilt	Leaves turn light in color and become bronzed from tips and margins inward. Wilting and dieback of branches occur with development of brown or black streaks in sapwood	No known cure. Girdle infected trees. Remove and burn as soon as possible
Pachysandra	Leaf and stem blight	Large brown to black areas on leaves with pink spore masses developing in wet weather	Remove diseased plants. Improve air circulation by thinning. In the spring during wet weather spray with mancozeb 80% WP 3 tbsp/gal. 3 applications at 10-day intervals
Periwinkle (vinca or ground myrtle)	Rhizoctonia Stem canker	Stems turn brown to black at soil level and shoots die out in patches	Spray weekly with any fixed copper fungicide until disease stops spreading
Pine	Needle cast	Occurs on Scotch, Austrian and Red Pine only. In early spring previous year's needles turn brown and drop	Spray with maneb 80% WP 1½-2 tbsp/gal. from August–October

		by midsummer. Small, black fungus fruiting bodies are present on fallen needles	
	Diplodia tip blight	Dieback of branch tips. Small black fungus fruiting bodies occur at the base of diseased needles. Occurs on Austrian, Red, and Scotch Pines	Just before new growth starts in the spring spray with bordeaux mixture and repeat at 14-day intervals with 2 or more sprays
	Needle rust	Small, cream-colored pustules on needles. Red Pine is very susceptible	Destroy Goldenrod and asters in vicinity of valuable trees
Planetree		*See* Sycamore	
Plum (ornamental)	Black Knot	Elongate, rough black swellings on twigs and small branches	In late winter prune out and destroy as many Black Knots as possible. Apply a dormant spray or bordeaux mixture or benomyl 50% WP 1 tbsp/gal.
Poplar	Leaf rust (Melampsora)	Yellowish-orange, powdery pustules on lower leaf surfaces during early summer	Apply zineb 75% WP 2 tbsp/gal. or mancozeb 80% WP 2 tbsp/gal. or maneb 80% WP 2 tbsp/gal. Just before first rust pustules appear
	Cytospora canker	Common on Lombardy Poplar. Sunken areas in bark of branches and trunks. Circular to oval cankers develop which eventually girdle and kill infected branches	Avoid wounding. If possible prune out cankers on small branches. Control insects and fertilize to maintain tree vigor
	Powdery mildew	White moldy patches on leaves	If serious spray with wettable sulfur 4-6 tbsp/gal.
Privet	Anthracnose	Leaves turn pale green in color, wilt and shrivel and remain clinging to stem	In wet weather spray several times at 10-day intervals with ferbam 76% WP 2 tbsp/gal. or benomyl 50% WP 1 tbsp/gal.
Pyracantha (Firethorn)	Fireblight	New shoots appear as if scorched with fire	*See* Crabapple

TABLE 6.13 *(Continued)*

Host	Disease	Description	Control
	Scab	Brown spots on leaves and fruit	In spring at budbreak and 10-14 days later at petal fall spray with benomyl 50% WP 1 tbsp/gal.
Quince (Japanese)	Fireblight	*See* Crabapple	
Redbud	Cercospora and other leafspots		If serious spray at budbreak and 10-14 days later with any fixed copper fungicide at manufacturer's direction
	Botryosphaeria twig blight	Dieback of twigs. Discolored, sunken areas on twigs	Prune out and destroy cankered twigs during the dormant season
Rhododendron	Leaf spots and blights	Dead areas on leaves with small fungus fruiting bodies	After flowering spray with bordeaux mixture
	Dieback (Phytophthora)	Terminal buds and leaves turn brown, curl up and droop. Cankers occur on small branches	Prune out infected branches and twigs. After bloom spray with bordeaux mixture
	Wilt	Terminal leaves turn yellow and wilt. Stems turn brown at soil level and below. Plants die when roots rot	Avoid planting in wet areas. Remove dead plants. Adjust soil acidity to pH 4.0-4.5 by addition of acid peats
Serviceberry (Amelanchier or Juneberry)	Rust (Gymnosporangium)	Horned galls on undersides of leaves, and also on fruits causing leaves to drop. Alternate host is the southern white cedar	An application of zineb 1⅓ tbsp/gal. applied at mid-bloom helps to control the disease. Do not plant near southern white cedar
Snowberry (Symphoricarpos)	Spot anthracnose (Sphaceloma)	Small dark purple or black spots with gray centers on leaves in early spring	Before buds open in the spring spray with liquid lime-sulfur 5 tbsp/gal. Follow with ferbam 2½ tbsp/gal. at 7-10 day intervals until fruit forms
	Anthracnose (Glomerella)	Cinnamon-colored spots on leaves, which turn black and fall prematurely	Same as for Spot Anthracnose

	Berry rot (Alternaria)	Berries turn yellow to brown and develop a soft, watery rot	Same as for Spot Anthracnose
Spirea	Fireblight	Blossoms turn black and shoots die back from tips. Dead leaves remain attached to dead shoots	*See* Crabapple
	Leaf spots	Various shapes, sizes and colors	*See* Ash anthracnose
	Powdery mildew	In wet weather a white powdery mold covers young leaves and shoots with distortion of flowers	Spray or dust with wettable sulfur 4-6 tbsp/gal.
Spruce		*See* Pine	
Staghorn sumac	Powdery mildew	White mold on leaves during late summer and early fall	Spray or dust with wettable sulfur starting when mildew first appears
Sweet fern (Comptonia)	Rust	Long, thin, brown, thread-like spore horns extending from leaves	Do not grow near pine nurseries
Sycamore	Anthracnose and leaf spots	*See* Maple	
Willow	Cankers	Discolored sunken areas on twigs, branches or trunk resulting in girdling and death of infected parts	Prune out infected parts. Same measures as suggested for poplar cankers
	Leaf spots	Small, round to irregular, brown spots on leaves	Rake up and destroy fallen leaves in the fall. Fungicides not usually necessary
Yew	Phytophthora root and crown rot	Plants slowly decline and die. The problem is frequently associated with poor drainage	Fertilize to maintain good vigor. Improve drainage

tsp = teaspoonful. WP = wettable powder. tbsp = tablespoonful. 1 tbsp/gal. = 1 lb/100 gal.

Wettable Sulfur is elemental sulfur of fine particle size combined with a wetting agent, and is sometimes referred to as "Dry Wettable Sulfur" sold under trade names such as Hy-Tox Sulfur, Kolosul, Magnetic 95, and Micro Flotox Wettable sulfur. It is suggested primarily for the control of powdery mildew.

Zineb is a tan-colored wettable powder containing 70–80% active ingredient, and is sold under trade names such as Dithane Z-78, Ortho Zineb Wettable, Blightox 65-W, and Parzate Zineb Fungicide.

LAWN AND TURF DISEASE CONTROL

Many people are unaware of the scope and importance of the turfgrass industry. It has been said that you stand, walk, run, picnic, and play on it. The "IT" refers to the green grass under your feet which may be your lawn, the city or state park, the golf course, the athletic field, the roadside shoulder or the green carpet that beautifies many modern industrial plants. It may come as a surprise to many people that the value of the turfgrass industry in the U.S.A. exceeds the annual value of several of the major agricultural crops grown in the U.S.A.

In recent years, despite booming industrialization, the importance of improved turfgrasses has been emphasized by the national trend towards landscaped shopping centers, mushrooming suburban residential development, interstate highway construction, cemetery beautification, and expanding recreational facilities. All of these require the establishment and maintenance of improved turfgrass plantings. The extent of the turfgrass industry in Indiana has been summarized in a recent Turfgrass Industry evaluation for 1976–1977, as shown in Table 6.14. In Indiana there are approximately 300,000 acres of mowed turfgrass areas bordering Indiana highways. At Purdue University, in Lafayette, Indiana, there are 100 acres of land for athletics, 220 acres in the 36-hole golf course, 480 acres of turf surrounding the permanent buildings, 500 acres at the Purdue Airport and roadsides, 20 acres in sod production, and 10 acres for turf improvement research. This comprises 1330 acres of turfgrass maintained and regularly mowed on the campus of Purdue University.

Maintenance of turfgrass is a technical profession involving the use of fertilizers, herbicides, insecticides, and fungicides. The extent to which diseases of turfgrass contribute to maintenance costs has not been accurately determined, but they have to be considered as one of several factors in the efficient maintenance of improved turfgrass areas.

Like all other living forms, the turfgrasses that constitute the home lawn, the golf course greens, or the green carpet of city parks, are subject to the vagaries of a constantly changing environment. Consequently they

TABLE 6.14. TURFGRASS INDUSTRY EVALUATION IN INDIANA (1976-1977)

Type	Number of Units	Acres	Annual Maintenance Cost per Acre ($)	Total Annual Maintenance Cost ($)
Airfields	176	76,280	32.77	4,600,000
Cemeteries	1,200	11,240	315.00	3,500,000
Churches	3,600	4,285	250.75	1,100,000
City, county and state buildings	7,080	11,000	250.00	2,900,000
Colleges	62	19,000	250.00	4,600,000
Grade schools and high schools	2,500	17,000	640.00	3,200,000
Golf courses	354	40,000	480.00	4,700,000
Industrial sites	48,660	15,000	290.00	4,300,000
Military installations	37	9,000	120.00	1,800,000
Parks (city and state)	410	25,000	370.00	13,500,000
Residential property (home lawns)	1,214,000	303,000	130.00	39,800,000
Resorts, motels, camps	—	13,284	420.00	5,600,000
Roadsides, state and federal	9,640	289,000	6.52	1,900,000
Roadsides, county	15,900	79,000	8.00	600,000
Roadsides, primary	29,400	235,000	9.00	2,100,000
TOTAL		1,147,089		94,200,000

Data courtesy of Dr. William H. Daniel, Midwest Regional Turf Foundation, Purdue University, West Lafayette, Indiana.

frequently fall victim to a multiplicity of ills which may spoil their appearance temporarily, or destroy their usefulness permanently. All of the major diseases of turfgrasses are caused by parasitic molds or fungi. Diseases of improved turf, whether they are in the homeowner's lawn or in the carefully groomed greens of the local country club, can be destructive and a cause for unexpected concern and expense. Nothing is more disturbing to the homeowner who has spent time and money to establish an attractive home lawn, than to have a spring epidemic of *Helminthosporium* Melting-Out suddenly appear, destroying large areas of the prized lawn. For the professional greenskeeper, an epidemic of Dollarspot or Large Brown Patch on the Number 4 green invariably results in complaints and criticism from the membership.

The use of chemicals for the prevention of turfgrass diseases IS ONLY ONE WEAPON IN THE PREVENTION OF UNWELCOME DISEASES. Before considering the use of turf fungicides on lawns, fairways, putting greens, and other turfgrass areas, it is important to follow the Ten Commandments of Turfgrass Care sumarized below:

1. Select grass species and strains suited to local soil and weather conditions. Whenever possible use disease-resistant cultivars of bluegrasses.
2. For seeding always use mixtures of grass varieties since the dif-

ferent types vary in their susceptibility to turfgrass diseases. A lawn seeded with a mixture of grass species or types will usually have one type survive specific disease epidemics.

3. Do not "scalp" upright growing grasses such as Kentucky bluegrass, Merion and Delta bluegrass, or Red Fescue. Close cutting of these species may greatly increase the severity of some disease problems. Maintaining a height of 2–3 in. by frequent mowing is best for grasses of the bluegrass type. Prostrate or creeping grasses such as bentgrasses, zoysia, and bermudagrass may be maintained at a cutting height of ½ in. or less.

4. Maintain enough fertility to keep the grass growing vigorously, but avoid excessive nitrogen applications which favor soft lush growth. Some of the turfgrass diseases such as Large Brown Patch and *Helminthosporium* leafspot are favored by high nitrogen fertility. Fertility requirements should be determined by professional soil tests in qualified soil analysis laboratories, and on the advice of professional soil fertility specialists.

5. During periods of rapid growth or during periods when diseases are especially troublesome, usually in May and June, grass clippings should be caught and removed, since fungus spores for new disease infections may originate from this source.

6. Frequent light watering should always be avoided. Nightly sprinkling with water helps to disseminate the spores of disease-inciting fungi and does little to maintain adequate soil moisture. When grass starts to wilt in hot weather, it is best to soak the soil to a depth of 4–6 in. every 7–10 days.

7. Remove excess thatch in early spring or early autumn when ½ in. or more has accumulated, using rented vertical mowers or power rakes.

8. Follow suggested insect and weed control programs advised for the region.

9. Provide good surface and subsurface drainage, and fill in low spots where water may accumulate.

10. Mow grass frequently so that no more that one-fourth to one-third of the leaf area is removed at any one time.

In spite of careful application of these "Timely Tips for Turf," plant diseases all too frequently become potentially destructive problems. When this happens the only recourse is to consider the safe and effective use of fungicides approved for the control of turfgrass diseases.

Two approaches using chemicals for the prevention of turfgrass diseases are possible. The most satisfactory approach is a preventative program applied in anticipation of possible disease occurrence. For the homeowner, the so-called "fly-swatter" method involves the use of turf fungicides after the disease is present. In such cases the disease probably will have occurred before it is recognized and identified, and considerable damage may occur before the control is applied. Turfgrass diseases are difficult to eradicate once they are well established, and will recur whenever weather conditions are favorable for the causal agent.

Fungicides for the control of turfgrass diseases are sometimes unavailable from some garden supply stores, but usually may be purchased from nurseries, city park maintenance departments, golf course greenskeepers, or the manufacturer. Unfortunately, the turfgrass fungicides are known by a variety of trade names and common names, or by their chemical names. This can be confusing to the home lawn-fancier as well as to some professional greenskeepers. For this reason, the common names, trade names, and chemical names of fungicides approved for use on turfgrass are summarized in Table 6.15. Specific suggestions for fungicides for control of turfgrass diseases are summarized in Table 6.16.

HOUSEPLANT DISEASE CONTROL

Growing plants for fun has become a winter hobby for many people who enjoy the fascination of watching plants grow. During the long, dull winter months a houseplant with its shiny green leaves or its occasional brightly colored blossom is an omen of spring. To some people, a growing plant is taken for granted as part of the natural terrestrial landscape, while others have learned to appreciate the beauty and complexity of an endless variety of skillfully engineered biological creations of every imaginable form and color.

A warm, comfortable living room might be considered an ideal place for any growing plant. But this is not always true. Plants need sunshine, or a substitute light source, in order to carry out the complicated process of photosynthesis, which is the daily bread for all growing plants. For this reason alone, the average living room is not the most desirable place for satisfactory plant growth. Plants, like humans, must have food for survival, and depend upon minerals in the soil as a food source. This is sometimes hard for the houseplant to find when it has been struggling to survive in the same soil for many months. Water, too, is just as necessary for plants as it is for people and animals. For plants, water is the vehicle which carries minerals from the soil to all parts of the plant. Water is also the method the plant uses to distribute plant foods manufactured in the plant factory—the leaf.

For the homeowner who has lavished TLC (tender loving care) on prized houseplants, an understanding of the form and function of living plants will in most cases avoid frustration and disappointment with the "living room greenhouse project."

TABLE 6.15. TRADE NAMES, COMMON NAMES, AND CHEMICAL NAMES OF FUNGICIDES FOR TURFGRASSES

Trade Name	Common Name	Chemical Name
Acti-dione	Cycloheximide	3-[2-(3,5-dimethyl-2-oxocyclohexyl)-2-hydroxyethyl]-glutarimide
Acti-dione PM	Cycloheximide	
Acti-dione RZ	Cycloheximide + pentachloronitrobenzene (PCNB)	
Acti-dione TGF	Cycloheximide	
Acti-dione-thiram	Cycloheximide + tetramethylthiuram disulfide	
Bedfume (fumigant)		methyl bromide
Benlate	Benomyl	methyl 1-(butylcarbamoyl)-2-benzimidazolecarbamate
Best Turf Disease Control	chlorothalonil	tetrachloroisophthalonitrile
Bromex (fumigant2		methyl bromide
Brom-o-gas (fumigant)		methyl bromide
Bromosan (fumigant)	thiophanate + thiram	
Calo-Clor		calomel + corrosive sublimate
Calo-Gran		calomel + corrosive sublimate
Captan	captan	N-(trichloromethylthio)-4-cyclohexene-1,2-dicarboximide
Chipco Spot-Kleen	thiophanate-methyl	
Chipco Thiram 75	thiram	tetramethylthiuram disulfide
Cleary's 3336	thiophanate	diethyl 4,4-o-phenylenebis (3-thioallophanate)
D & P	thiram	tetramethylthiuramdisulfide
Daconil 2787	chlorothalonil	tetrachloroisophthalonitrile
Dexon	fenaminosulf	p-dimethylaminobenzenediazo sodium sulfonate
Dyrene	anilazine	2-4-dichloro-6-(o-chloroanilino)-s-triazine
Ethisul	—	polyethylenethiuramsulfide
Fore	mancozeb	coordination product of zinc ion and manganous ethylenebisdithiocarbamate
Fungiclor	PCNB	pentachloronitrobenzene
Fungo	thiophanate-methyl	Dimethyl 4,4-o-phenylenebis (3-thioallophanate)
Iscobrome	methyl bromide	
Koban	etridiazole	5-ethoxy-3-trichloromethyl-1,2,4-thiodazole

MBC Fumigant	methyl bromide	
MC-2	Dowfume	methyl bromide
Mertect	thiabendazole	2-(4-thiazolyl) benzimidazole
Micofume	DMTT	3,5-dimethyl tetrahydro-1,3,5-2H thiadiazine-2-thione
Mylone	dazomet	tetrahydro-3,5-dimethyl-2H-1,3,5-thiadiazine-2-thione
Orthocide	captan	N-(trichloromethylthio)-4-cyclohexene-1,2-dicarboximide
Ortho Lawn Disease Control		Captan + PCNB
Pestmaster Soil Fumigant	methyl bromide	
Phaltan	folpet	N-(trichloromethylthio)phthalimide
Plantvax	oxycarboxin	5,6-Dihydro-2-methyl-1,4-oxothiin-3-carboxanilide-4,4-dioxide
Proturf DSB Fungicide	benomyl	methyl 1-(butylcarbamoyl)-2-benzimidazolecarbamate
Proturf Fungicide *II*	PCNB	pentachloronitrobenzene
Proturf Fungicide *III*	anilazine	2-4-dichloro-6-(o-chloroanilino)-s-triazine
Soil Fumigant-M	DMTT	3,5-dimethyl tetrahydro-1,3,5-2H thiadiazine-2-thione
Spectro		anilazine + thiophanate
Spot-Trete	thiram	tetramethylthiuramdisulfide
Terraclor	PCNB	pentachloronitrobenzene
Terrazole	ethazol	5-ethoxy-3-trichloromethyl
Tersan 75	thiram	tetramethylthiuramdisulfide
Tersan 1991	benomyl	methyl 1-(butylcarbamoyl)-2-benzimidazolecarbamate
Tersan LSR	maneb	manganous ethylenebisdithiocarbamate
Tersan SP	chloroneb	1,4-dichloro-2,5-dimethoxybenzene
Thiram 75	thiram	tetramethylthiuramdisulfide
Thiramad	thiram	tetramethylthiuramdisulfide
Thiuram	thiram	tetramethylthiuramdisulfide
Tobaz	thiabenzadole	2-(4-thiazoyl) benzimidazole
Topsin	thiophanate-methyl	Dimethyl 4,4-o-phenylenebis(3-thioallophanate)
Vapam	SMDC	sodium methyldithiocarbamate
Vorlex	MIT	methyl isothiocyanate

TABLE 6.16. CHEMICAL CONTROL OF TURFGRASS DISEASES

Disease	Causal Agent	Susceptible Grasses	Fungicide and Rate/1000 sq ft	Comments
Algae	—	All turfgrasses	Copper sulfate (1-2 oz) or Fore WP 80% (2-4 oz) or Tersan LSR WP 80 (2-4 oz)	Under wet conditions repeat as needed
Brown Patch	*Rhizoctonia solani*	Bentgrass Bermudagrass Bluegrass Fescue Ryegrass Zoysia	Acti-dione-thiram (2-4 oz) or Daconil 2787 WP 75 (4 oz) or Dyrene WP 50 (4-6 oz) or Fore WP 80 (4-6 oz) or Fungo WP 50 (6 oz) or Manzate 200 WP 80 (4-6 oz) or Tersan 1991 WP 50 (2-4 oz) or Tersan LSR WP 80 (4-6 oz)	Avoid high nitrate fertilizers. Improve soil aeration and drainage
Copper Spot	*Gloeocercospora sorghi*	Bentgrass	Cleary 3336 WP 50 (1 oz) or Daconil 2787 WP 75 (2-4 oz) or Dyrene WP 50 (4-6 oz) or Fungo WP 50 (6 oz)	Likely to occur June-October at maximum temperatures of 68°–75°F
Dollar Spot	*Sclerotinia homeocarpa*	Bentgrass Bermudagrass Fescue Ryegrass Zoysia	Acti-dione-thiram (2-4 oz) or Cleary 3336 WP 50 (1 oz) or Daconil 2787 WP 75 (2-4 oz) or Dyrene WP 50 (2-4 oz) or Fungo WP 50 (1 oz) or Mertect WP 60 (2 oz) or Tersan 1991 WP 50 (2 oz)	May-November reapply if necessary at 7-10 day intervals. Apply adequate nitrogen
Fairy Ring	Several toadstool fungi	All turfgrasses	To eradicate, fumigate with methyl bromide	Remove infested soil and replace with clean soil. Reseed or resod
Fusarium Blight	*Fusarium roseum*	Bentgrass Bluegrass Fescue	Tersan 1991 WP 50 (6-8 oz) or Cleary 3336 WP 50 (1 oz) or Fungo WP 50 (1 oz)	Water thoroughly the day before treatment. Avoid heavy nitrogen

[1]Bluegrasses refers to both annual bluegrass (*Poa annua*) and Kentucky Bluegrass (*Poa pratensis*).

		Ryegrass		in spring. Must be drenched with water after application
Fusarium Patch (Pink snow mold)	*Fusarium nivale*	All turfgrasses	Dyrene WP 50 (4-6 oz) or Fore WP 80 (4-6 oz) or Tersan 1991 WP 50 (2 oz) or Tersan SP WP 65 (6-9 oz)	Late fall to early spring. Avoid fall nitrogen
Grease Spot (Pythium Blight)	*Pythium aphanidermum* *P. ultimum*	All turfgrasses	Koban WP 35 (4 oz) or Tersan SP WP 50 Manufacturer's recommendation	July-August. Improve soil drainage. Use sufficient water in fungicide application for adequate penetration
Helminthosporium Leafspot	*H. sorokiniarum*	Bentgrass Bluegrass Fescue Ryegrass	Acti-dione-thiram (2-4 oz) or Captan WP 50 (4-6 oz) or Daconil 2787 WP 75 (4 oz) or Dyrene WP 50 (4-6 oz) or Fore WP 80 (4-6 oz) or Manzate 200 WP 80 (4-6 oz) or Tersan LSR WP 80 (4-6 oz) or Zineb WP 75 (4-6 oz)	June-August. Remove clippings. Raise cutting height. Fertilize to maintain vigor
Helminthosporium Melting-out	*H. vagans*	Same as Heminthosporium Leafspot		
Moss	—	All turfgrasses	Apply ferrous ammonium sulfate (16 oz)	
Powdery mildew	*Erysiphe graminis*	Bermudagrass Bluegrass Fescue	Acti-dione-thiram (2-4 oz) or Acti-dione TGF (2 oz) or Tersan 1991 WP 50 (2-3 oz)	July-September. Repeat at 7–14 day intervals as needed
Pythium blight	*See* Grease Spot			
Red Thread	*Corticium fuciforme*	All turfgrasses Mostly fescue, ryegrass	Daconil 2787 WP 75 (4 oz) or Cleary's 3336 WP 50 (1 oz) or Fungo WP 50 (2 oz)	Increase nitrogen

TABLE 6.16 *(Continued)*

Disease	Causal Agent	Susceptible Grasses	Fungicide and Rate/1000 sq ft	Comments
Rust	*Puccinia graminis f.sp. agrostis*	Bluegrass especially Merion Ryegrass Zoysia, especially Manhattan	Acti-dione-thiram (4 oz) or Daconil 2787 WP: 75 (4 oz) or Dyrene WP 50 (4-6 oz) or Fore WP 80 (4 oz) or Tersan LSR WP 80 (3-4 oz) or Zineb WP 75	June-September. Treat every 7–14 days until rust disappears
Seed rots, damping-off, seedling blights	Several species of fungi	All turfgrasses	Seed treatment with Captan WP 50 or Thiram WP 75 Seedling protection Captan WP 50 (2-4 oz) or Daconil 2787 WP 75 (2-4 oz) or Folpet (Phaltan) WP 50 (2 oz) or Zineb WP 75 (2-4 oz)	
Slime molds	*Physarum cinereum, Fuligo* spp.	All turfgrasses	Fore WP 80 (6-8 oz) or Zineb WP 75 (2-4 oz)	Remove mechanically by raking, mowing, or hosing with strong water stream
Smut (Flag)	*Urocystis agropyrii*	Bentgrass Bluegrass especially Merion Meadow Fescue Red Top Ryegrass	Fungo WP 50 (4-8 oz) or Tersan 1991 WP 50 (6 oz) or Cleary 3336 WP 50 (1 oz)	Apply in early spring before growth starts or in late fall. After application, drench with water for root zone penetration
Smut (Stripe)	*Ustilago striiformis*		Same as Flag Smut	
Typhula Blight (gray snowmold)	*Typhula itoana*	Bentgrass Bluegrass (under snow)	Tersan SP or Dyrene or Tersan 1999 at manufacturer's direction	

WP = wettable powder. 1 oz/1000 sq ft = 5½ tbsp/1000 sq ft.

All plants are composed of cells of many types, shapes, and sizes, and each cell has a special function. The general structure of a plant is illustrated in Fig. 6.1. The underground plant parts are the roots, which serve the plant by absorbing water and necessary minerals from the soil. They also support the plant and hold it upright. The cells in the plant's roots are similar in arrangement to those in the stem, and contain xylem cells (water-conducting tissues) and phloem vessels (food-transporting tissues). In the stem the xylem tissues also transport minerals and water absorbed by the roots, and the phloem in the stem carries food material to various parts of the plant. The stem also serves the very useful purpose of supporting the leaves in a position where they will receive the greatest amounts of light.

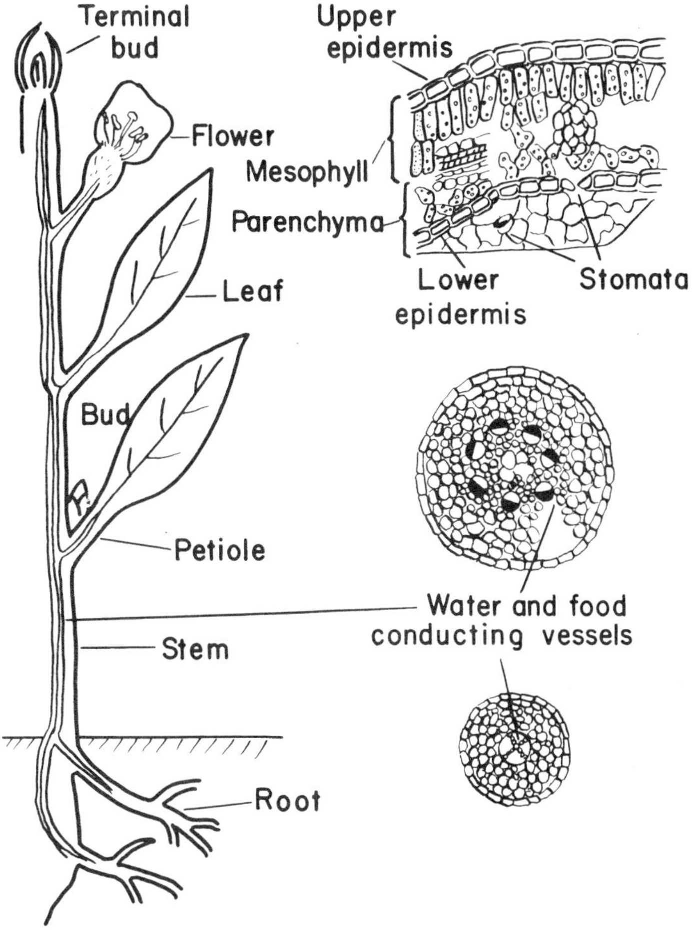

FIG. 6.1. THE STRUCTURE OF PLANTS

The leaf is the plant factory, and is a thin expanded structure having a large surface which permits a great number of the cells to be exposed to light. The leaf cells contain "chlorophyll," a complex green pigment that gives the leaf its color. The chlorophyll in leaves utilizes energy from the sun's rays and changes it into a form of energy that can be used by the plant in the manufacturing of plant foods (sugars). The plant leaf absorbs carbon dioxide from the air and releases oxygen through special pores known as "stomata." This interesting and complicated process is known as "photosynthesis" and may be illustrated by the following simple equation:

$$\text{Carbon dioxide} + \text{water} \xrightarrow[\text{and sunlight produces}]{\text{with the aid of chlorophyll}} \text{Sugars} + \text{Oxygen}$$

The detailed equation for photosynthesis is usually written:

$$6\,CO_2 + 6H_2O + 672\text{ Kcal} \rightarrow C_6H_{12}O_6 + 6\,O_2$$

Carbon dioxide	water	radiant energy	Glucose	oxygen

One Kilocalorie (Kcal) is the energy required to heat 1000 g water 1°C, or is the energy required to heat 1 qt water 2°F. Burning 1 g mol/wt (180 g or 6 oz) glucose yields 672 Kcal. This is sufficient to heat enough water for a good hot shower.

The process of photosynthesis occurs in two phases: a photo phase and a synthetic phase, as illustrated in Fig. 6.2.

If all the conditions for satisfactory plant growth are in balance and normal photosynthesis proceeds, plants will thrive unless they are attacked by pests and diseases, or adversely affected by mechanical injury or unfavorable environmental conditions. It is rare for all factors to be favorable for optimum plant growth, especially in the average home during the winter months. One of the major problems in the maintenance of houseplants is determining how to supply all the plants' needs for growing in unnatural surroundings. To enjoy success with houseplants, the homeowner must take the time to learn about the requirements, since in many instances sick houseplants frequently suffer from lack of adequate light, too much water, too dry a climate, improper soil, or excess of or lack of adequate fertility.

In spite of liberal doses of TLC, houseplants may frequently fall victim to plant diseases such as root rots, damping-off, vascular wilts, and leafspot diseases. Prescriptions for the prevention of common houseplant problems are outlined in Table 6.17.

TABLE 6.17. PRESCRIPTIONS FOR THE PREVENTION OF HOUSEPLANT PROBLEMS

Plant	Disease	Symptoms	Control Suggestions
Seedlings	Damping-off	Seedlings turn dark in color at soil level; they eventually rot and fall over	Use sterilized soil; dust seed with captan; dust surface of soil with ferbam or captan; drench soil with captan 1 tbsp/gal. water
Most leafy ornamentals	Powdery mildew (F)	White powdery growth on surface of leaves, buds and stems	Provide better ventilation; avoid wetting leaves with overhead watering systems. Apply spray of Karathane 2 tsp/gal. water
Most houseplants	Nematodes	Plants stunted and sickly; swellings usually occur on roots	Make new leaf cuttings in peat or vermiculite; plant in sterilized soil; discard infested plants
African Violet	Crown Rot and Foot Rot (F)	Plants sickly; wilting occurs with development of a brown soft-rot of crowns and leaf stems	Avoid over-watering; drench soil with ferbam 2 tbsp/gal. water
	Botrytis blight (Gray mold) (F)	Soft, tan-colored rot of crowns, buds and flowers	Increase air circulation, avoid wetting leaves; spray with zineb 1 tbsp/gal. water at 7-10 day intervals
	Leaf scorch (NP)	Leaves yellowish, scorched at margins and tips, and may wilt	Shade plants from direct sun
	Mosaic (V)	Plants dwarfed with crinkled leaves; leaves mottled light and dark green between the veins	Destroy infected plants; start new virus-free plants and control insects such as aphids, white fly and spider mite

TABLE 6.17 *(Continued)*

Plant	Disease	Symptoms	Control Suggestions
Asparagus fern	Leaf drop (NP)	Leaves turn brown, wither and drop	Room temperature too high—lower to 70-72°F and increase humidity
Begonia	Botrytis blight (Gray mold) (F)	Soft, tan-colored rot of crowns, buds, flowers and leaf stems	Increase air circulation; avoid wetting foliage by sprinkling; spray with zineb 1 tbsp/gal. water at 7-10 day intervals
	Stem, crown and root rot (F)	Stems and crowns water-soaked and darkened at soil level; stems collapse and die	Discard infected plants; drench soil with ferbam or zineb, 1 tbsp/gal. water
	Root knot	Caused by nematodes, small swellings on roots	Make new leaf cuttings in peat or vermiculite; plant in sterilized soil, discard infested plants
	Mildew (F)	White powdery growth on leaves	Provide better ventilation; avoid sprinkling leaves with water; spray with Karathane 1 tsp/gal. water
Cactus	Root rot (F)	Roots and base of plant turn soft and rot	Repot in sterilized soil; water sparingly and uniformly; place in larger pot
Calla Lily	Bulb and root rot (F)	Plants stunted, sickly and wilted; yellow leaves; bulbs and roots rotten and decayed	Plant only disease-free bulbs; provide good drainage in pot; remove infected plants; dip bulb and base of plant in 1-1000 bichloride of mercury and repot in sterilized soil
Chinese Evergreen (Philodendron,	Anthracnose (F)	Dead spots on leaves; restricted	Avoid chilling; do not sprinkle leaves when watering; spray

Dieffenbachia, Agloenema)		brown spots which may run together to form extensive dead areas	with zineb 1 tbsp/gal. water at 7-10 day intervals
Ferns	Botrytis blight (Gray mold) (F)	Soft, tan-colored rot of crowns, leaves and stems	Increase air circulation; avoid sprinkling leaves; spray with zineb 1 tbsp/gal. water at 7-10 day intervals
	Brown leaf galls	These are not a sign of disease but rather natural spore-producing receptacles	
Gardenia	Chlorosis (NP)	Leaves yellow and sickly	Caused by iron deficiency; treat with Sequestrene at manufacturer's direction, or add a pinch of iron sulfate to soil
	Bud drop (NP)	Buds fall from plants	Avoid chilling or sudden quick change of temperature; maintain high humidity
	Stem canker (F)	Sunken, discolored areas on stems at soil level	Avoid over-watering and poor drainage; drench soil with bordeaux mixture or any other copper fungicide
Geranium	Botrytis blight (Gray mold) (F)	Soft, tan-colored rot of crowns, buds, flowers and leaf stems	Increase air circulation; avoid sprinkling leaves; spray with zineb 1 tbsp/gal. water
	Stem rot (B)	Soft, dark, slimy rot at soil level	Make new leaf cuttings; plant in sterilized soil, destroy diseased plants
	Dropsy (Edema) (NP)	Water-soaked spots on leaves, later turning corky; corky ridges on leaf stems	Avoid high soil moisture; provide better light and ventilation
Palm	Anthracnose (F)	Water-soaked spots on leaves which later turn brown	Remove and destroy infected leaves; spray with zineb 1½ tbsp/gal. water

TABLE 6.17 *(Continued)*

Plant	Disease	Symptoms	Control Suggestions
		and fall out, leaving holes in leaves	
Peperomia	Ring spot (V)	Distorted leaves with striking concentric zonal spots	Make new leaf cuttings; plant in sterilized soil; discard infected plants
Poinsettia	Chlorosis (NP)	Leaves turn yellow and drop	Avoid chilling and over-watering and provide adequate lighting
	Bacterial Canker (B)	Long, water-soaked streaks on one side of green stems	Make new cuttings and plant in sterilized soil; avoid overhead watering; destroy infected plants
	Stem rot (F)	Lower portions of stem turn black and decay	Start new cuttings from disease-free plants in sterilized sand; drench soil with mixture of 1 tbsp Terraclor plus $1\frac{1}{4}$ tbsp ferbam per gallon of water
Rubber plant (*Ficus elastica*)	Anthracnose (F)	Dark-colored, restricted spots on leaves; dead tissue later dries and falls out	Spray with zineb $1\frac{1}{2}$ tbsp/gal. water

tsp = teaspoonful. tbsp = tablespoonful. (F) = caused by a fungus. (B) = caused by a bacterium. (V) = caused by a virus. (NP) = non-parasitic disease.

184 PLANT DISEASE CONTROL

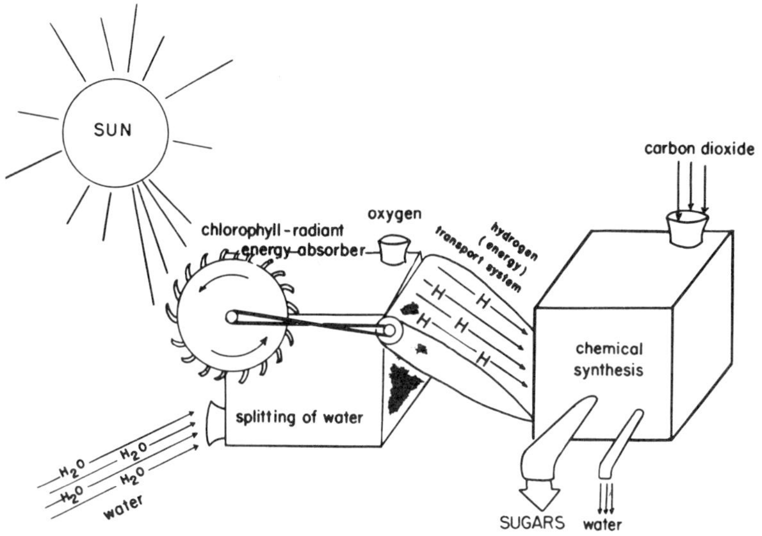

Adapted from Calvin

FIG. 6.2. THE PROCESS OF PHOTOSYNTHESIS

HOUSEHOLD MILDEW AND WOOD ROT CONTROL

Many a homeowner, motel or hotel operator, or lake resort owner is familiar with household mildews and wood rots. It is frequently a surprise when shower stalls, basement walls, shoes, books, golfclub bags, and canvas items suddenly become covered with annoying, persistent, and disfiguring growth of mold or mildew. Discoloration of shower stalls, painted basement walls, and deterioration of wooden floors, decks, and porches at the lake cottage are painful reminders that while molds, or fungi, have not inherited the earth, they do in some cases take their toll of goods and products as well as plants. The activities of these microscopic molds have been described in an interesting book by C.M. Christenses (1951) entitled "The Molds and Man."

Seymour S. Block, Professor of Chemical Engineering at the University of Florida, Gainesville, in the introduction of his chapter on Industrial Preservatives in Torgeson's Fungicide Treatise (1967), stated:

> Decay and rebirth may be nature's scheme, but they do not always satisfy man's desires, even in this age of plenty and planned opulence. Fungi (molds) are in the forefront of the war of degradation. They show no more respect for the cotton in a canvas awning sheltering the V.I.P.'s at the Waldorf Astoria than for the cotton in the boll in a field in Texas. They are at work in missile components, plastic coatings, in television sets, pharmaceutical products and paper mills.

One of the reasons for the use of synthetic fabrics in the garment manufacturing industry is the susceptibility of cotton fiber to invasion by molds, resulting in discoloration. Clothing stored in warm damp places is subject to "mildew," especially in tropical and sub-tropical regions or when stored for long periods of time under conditions of high humidity. Leather goods such as shoes and suitcases waiting in the attic or basement for the next vacation trip, and leather bound books and other leather items may become covered with mildew resulting in permanent staining or serious deterioration.

Molds may also break down electrical insulation, and can etch the lenses of cameras, binoculars, and telescopes. About the only item that molds cannot affect are metal products. Fungi, or molds, can exist on casein glues used in the fabrication of plywood panels, and under moist conditions, when in association with bacteria, can slowly digest and weaken the glue resulting in warped panelling or separation of the layers of veneer forming the panelling.

Suggestions for the prevention and treatment of household molds are summarized in Table 6.18.

Wood Rots

There are many species of molds or fungi that can live exclusively on wood, causing rot and ultimate decay. In general, the lighter colored woods such as basswood and the true firs will decay more rapidly than darker woods such as white oak, western red cedar, and walnut. Under conditions favorable for the growth of fungi all types of wood can be attacked by wood-rotting fungi.

There are many types of fungus-incited wood rots in living trees which may shorten the trees' lives if they are not treated correctly before major damage has resulted. Wood rots of fence posts, greenhouse benches, wooden storage structures, window frames, wooden steps, roof sheathing, and rotting of structural timbers of residences and other wooden buildings are examples of wood rots caused by fungi. Fungus wood rots are greatly influenced by temperature, moisture, and other factors. In general, whenever wood is wet but not submerged in water, it will be attacked by wood-rotting species of fungi. Green lumber used in construction with a moisture content of 50% or above will take months to dry out. Green lumber of this type, when used in construction work for framing and sheathing of roofs and walls, will eventually develop wood rot, resulting in serious problems long after the construction has been completed. On the other hand, well seasoned lumber with a moisture content of 25–28% generally will remain free from wood rot problems. Wood that is completely submerged in water will rarely decay

TABLE 6.18. PREVENTION AND TREATMENT OF HOUSEHOLD MILDEW

Item Concerned	Suggested Treatment
General Suggestions for the Prevention of Mildew	Since the molds or fungi responsible for mildew on common household items flourish in moist places, anything that can be done to lower the moisture content of the air will help to reduce the danger of "mildewing." Adequate ventilation to provide good air circulation, and the use of electric dehumidifiers and silica gel dessicants are good preventative measures for avoiding household mildew
Cement blocks, basement walls, and partitions	Wash thoroughly with a laundry bleach solution (sodium hypochlorite) containing ½-1 c bleach per gallon of warm water. Rinse with fresh water after treatment
Clothing and fabrics	Moisten the mildew stain with lemon juice, and sprinkle the affected area with salt and place outdoors in the sunshine or soak the stained area in a solution of 4 tbsp sodium perborate/ 1 pt lukewarm water and rinse in clean water after treatment
Moldy leather goods	Sponge the affected area with rubbing alcohol containing 1% aspirin (Salicylanilide), or dichlorophene, thymol, or hexachlorophene. After treatment protect with a coating of floor wax containing mildew-proofing chemical (paranitrophenol or hexachlorophene). For mildew-contaminated shoes wash the insides of the shoes with germicidal soap (Dial, Lifebuoy), and swab with formaldehyde and store for 1 hour in a closed plastic bag
Books and paper products	Swab affected area lightly with rubbing alcohol and protect with any available artist's fixative such as "Fixatif Eagle Art Spray," manufactured by Eagle Pencil, a division of Berol Corporation, or similar product
Articles stored in closets	Mothballs or crystals of paradichlorobenzene used at the rate of 2 oz/100 sq ft storage space will effectively prevent mildew
Washable textiles (cotton curtains, awnings, tents, canvas goods, etc.)	Dip affected item in hot neutral soapsuds for a few minutes, and then soak in hot copper sulfate solution (Blue Vitriol or Bluestone) used at the rate of 1½ oz/gal. water. Soak for 15 min with constant stirring
	OR
	Dip in or brush with a fungicide containing Copper Naphthenate (Cuprinol, Copper-Cure, Rot-Not, Wittox 'C', etc.) diluted according to manufacturer's directions in diesel fuel or petroleum oil. Since these products are green in color they may impart a green color to treated awnings, tents and canvas goods. If this is objectionable use the copper sulfate (Blue Vitriol) treatment

because most of the wood-rotting fungi require oxygen for their growth and development.

BIBLIOGRAPHY

Useful References on Vegetable Diseases

BOSWELL, V.B., S.P. DOOLITTLE, and L.M. PULZ. 1951. Pepper production, disease and insect control. U.S. Dep. Agric. Farmer's Bull. *2051*. Washington, D.C.

BROGDON, J.E., M.E. MARVEL, and R.S. MULLIN. 1960. Commercial vegetable insect and disease control. Cir. 193A. Fla. Agric. Exp. Stn., Gainesville.

CHUPP, C., and R.W. LEIBY. 1931. The control of diseases and insects affecting vegetable crops. Ext. Bull. *206*. Cornell Univ. Agric. Exp. Stn., Ithaca, N.Y.

CHUPP, C., and A.F. SHERF. 1960. Vegetable Diseases and Their Control. The Ronald Press, New York.

HILDEBRAND, E.M., and H.T. COOK. 1959. Sweet potato diseases. U.S. Dep. Agric. Farmer's Bull. *1059*. Washington, D.C.

LINN, M.B. 1958. Vegetable diseases. Univ. of Illinois. Agric. Ext. Ser. Cir. 802. Urbana.

MUNCIE, J.H. 1943. Michigan potato diseases and their control. Ext. Bull. *162*. Mich. Agric. Ext. Ser., East Lansing.

REED, L.B., and R.E. WEBB. 1967. Insects and diseases of vegetables in the home garden. U.S. Dep. Agric. Home & Garden Bull. *46*. Washington, D.C.

WALKER, J.C. 1957. Plant Pathology. McGraw-Hill Book Co., New York.

WALKER, J.C., and R.H. LARSON. 1961. Onion diseases and their control. U.S. Dep. Agric., Agric. Handb. 208. Washington, D.C.

WALKER, J.C., R.H. LARSON, and A.L. TAYLOR. 1958. Diseases of cabbage and related plants. U.S. Dep. Agric., Agric. Handb. 144. Washington, D.C.

WEBB, R.E., J.M. GOOD, and L.L. DANIELSON. 1967. Tomato diseases and their control. U.S. Dep. Agric., Agric. Handb. 203. Washington, D.C.

WHITAKER, T.W., and E.J. RYDER. 1962. Lettuce and its production. U.S. Dep. Agric., Agric. Handb. 221. Washington, D.C.

ZAUMEYER, W.J., and H.R. THOMAS. 1958. Bean diseases and their control. U.S. Dep. Agric. Farmer's Bull. *1692*. Washington, D.C.

Useful References on Fruit Diseases

ANDERSON, H.W. 1956. Diseases of Fruit Crops. McGraw-Hill Book Co., New York.

JONES, A.L. 1971. Diseases of Tree Fruits in Michigan. Mich. Agric. Exp. Stn. Ext. Bull. *E-714*. East Lansing.

KLOS, E.J. 1960. Small fruit diseases in Michigan. Mich. Agric. Exp. Stn. Ext. Bull. *370*. East Lansing.

KLOS, E.J. 1960. Tree fruit diseases in Michigan. Mich. Agric. Exp. Stn. Ext. Bull. *361*. East Lansing.

PECKNOLD, P. *et al.* 1979. Indiana 1979 fruit spray schedules. Purdue Agric. Exp. Stn. Coop. Ext. Ser. Publ. ID 60. Lafayette.

POWELL, D. *et al.* 1965. Diseases of apple and pears in the Midwest. North Central Regional Ext. Publ. 16. Ill. Agric. Exp. Stn. Coop. Ext. Ser. Cir. 909. Urbana.

Useful References on Flower Diseases and Their Control

FORSBERG, J.L. 1963. Diseases of ornamental plants. Univ. of Illinois Spec. Publ. 3. Urbana.

HORNE, W.C., and H.E. SMITH. (ND) What you should know about plant diseases. Texas A & M Univ. Agric. Ext. Ser. Publ. B. 995. College Station, Texas.

MCCAIN, A.H. 1970. Chemicals for plant disease control. Univ. of California Agric. Exp. Stn. Publ. MA-28. Berkeley.

MCCAIN, A.H. 1975. Fungicides for ornamental plants. Univ. of California Agric. Exp. Stn. Leaflet 2284. Berkeley.

MOLLER, W.J., D.H. HALL, A.H. MCCAIN, and A.O. PAULUS. 1975. Study guide for agricultural pest control advisers on plant diseases. Univ. of California Publ. 4046. Berkeley.

PIRONE, P.P. 1970. Diseases and Pests of Ornamental Plants. The Ronald Press, New York.

POWELL, C.C., JR. 1976. Disease control in the landscape. Ohio Agric. Exp. Stn. Bull. *614*. Columbus.

POWELL, C.C., JR., and R.K. LUNDQUIST. 1976. Insect, mite, and disease control on commercial floral crops. Ohio Agric. Exp. Stn. Bull. *538*. Columbus.

SHURTLEFF, M.C. 1962. How to Control Plant Diseases in the Home and Garden. Iowa State University Press, Ames.

THOMSON, W.T. 1977. Tree, turf, and ornamental pesticide guide. Thomson Publications. P.O. Box 7967, Fresno, Calif.

Useful References on Diseases of Shadetrees, Shrubs and Vines

CARTER, J.C. 1964. Illinois trees and their diseases. Illinois State Natural History Survey Cir. 45. Urbana.

MARSHALL, R.P., and A. WATERMAN. 1948. Common diseases of important shade trees. U.S. Dep. Agric. Farmer's Bull. *1987*.

NICHOLS, L.P. 1955. Diseases of ornamental shrubs and vines. Pennsylvania State Univ. Coop. Ext. Ser. Cir. 429. University Park.

PIRONE, P.P., B.O. DODGE, and H.W. RICKETT. 1970. Diseases and Pests of Ornamental Plants. 4th edition. The Ronald Press, New York.

POWELL, C., JR. 1976. Disease control in the landscape. Ohio State Univ. Coop. Ext. Ser. Bull. *614*. Columbus.

POWELL, C.C., JR., and M.J.A. BROWN. Spraying chemicals to control diseases of trees and shrubs. Current uses and their relationship to registration and labelling. Ohio State Univ. Dep. Plant Pathol. (unnumbered mimeographs), Columbus.

Useful References on Houseplant Problems

BICKERT, H.M., M.R. HARRISON, S.N. DAVIS, and J.L. LIBBY. 1961. Care of houseplants. New Jersey Agric. Exp. Stn. Coop. Ext. Ser. Ext. Bull. *337*. Princeton.

CATHEY, H.M., H.H. KLEUTER, and W.A. BAILEY. 1967. Indoor gardens for decorative plants. U.S. Dep. Agric. Home & Garden Bull. *133*. Supt. of Documents. Washington, D.C.

CHRISTENSES, C.M. 1951. The Molds and Man. University of Minnesota Press, Minneapolis.

FORSBERG, J.L. 1963. Diseases of ornamental plants. Univ. of Illinois Coop. Ext. Ser. Spec. Publ. 3. Urbana.

KIRBY, R.S., O.D. BURKE, and L.P. NICHOLS. 1955. Diseases of commercial florists crops. Pennsylvania State Univ. Coop. Ext. Ser. Cir. 423. University Park.

PIRONE, P.P., B.O. DODGE, and H.W. RICKETT. 1970. Diseases and Pests of Ornamental Plants. 4th edition. The Ronald Press, New York.

SHURTLEFF, M.C. 1967. How to Control Plant Diseases. Iowa State University Press, Ames.

TORGESON, D.C. 1967. Fungicides. Academic Press, New York.

WESTCOTT, C. 1950. Plant Disease Handbook. Van Nostrand Co., Princeton, N.J.

7

Chemical Control of Plant Diseases by Seed Protection

Accidental discovery has frequently been the source from which has stemmed many major contributions to modern science and technology. Thus it was with chemical seed treatment for plant disease control. As previously noted in Chapter 4, in the year 1670 a sailing vessel loaded with wheat encountered a storm in the Bristol Channel, England, ran aground, and foundered. Farmers along the coast salvaged some of the grain, and, discovering that it was too saturated with salt water to be used for food, planted some of it as seed. When the wheat from this seed grew and produced a harvest, it was noticed that the grain was relatively free from stinking smut disease. Wheat in nearby fields grown from conventional seed was heavily infested with the grain-destroying stinking smut disease. It was in this way that the practice of soaking seed wheat in salt brine for the control of stinking smut originated and was practiced by English farmers of the 17th and 18th Centuries.

In 1637 Remnant had reported the value of sodium chloride (salt) as a seed treatment of wheat for the prevention of stinking smut or bunt disease (*Ustilago tritici*). In 1733 Jethro Tull published in the London Times a lengthy description of a salt-brine treatment for stinking smut of wheat, and introduced the practice into practical agricultural usage in England. A little later, in 1760, Schulthess suggested the use of copper sulfate for the same purpose.

In 1755 Matthieu Tillet, a French botanist, had suggested the use of lye as a seed treatment for wheat. He recommended washing the seed with water, drying it, then soaking it in a solution of lye, drying it, and finally sprinkling it with a lye solution followed by a light dusting of lime. The results of his experiments so impressed Louis XV that he ordered the printing of these directions and their distribution throughout France at government expense. (See p. 51 for additional information about du Tillet and his contribution.)

In 1807 Bénédict Prévost, a Swiss Professor of Philosophy at the Academy at Montauban, France, published his classical "Memoir on the Immediate Cause of Bunt or Smut of Wheat, and of Several Other Diseases of Plants, and on Preventives of Bunt." In this memoir, Prévost presented an accurate description of the disease, described spore germination, and confirmed Tillet's observation that smut spores placed on the surface of wheat seed resulted in infection of the developing wheat plants. He also demonstrated that solutions containing copper sulfate prevented spore germination and distinguished between fungicidal and fungistatic effects.

"Danish Hot Water" was discovered and introduced by Jens Ludwig Jensen. Jensen was a Danish schoolmaster who edited and distributed an agricultural magazine during the period 1868-1880. In 1881 he organized his own private agricultural research institute, which he called "Bureau Ceres," in the city of Copenhagen. He was aware of the serious nature of the late blight disease of potatoes and published privately a pamphlet, "How to Overcome the Potato Disease." He emphasized the need to eliminate the source from which the disease originated, and asked the question "What was there then under Heaven, that could be put into a potato to kill a tangle of fungal threads interwoven among the cells, without harming the potato itself?" He suggested the answer to the riddle was heat, and thus pioneered thermal sterilization for the elimination of phytopathogens. In 1887, he published his method for the eradication of *Phytophthora infestans* (the potato late blight pathogen), in the "Mémoirs de la Société Nationale d'Agriculture" of France, and received the Sociétié's Gold Medal for his contribution. In this publication he suggested the use of dry heat, exposing infected tubers to a temperature of 40°C for 4 hours, and prescribed a simple apparatus to expedite the treatment.

It is not generally known that the "Hot Water Treatment" for elimination of loose smut of wheat followed the demonstration by Jensen of the practicability of thermal treatment for the eradication of the potato late blight fungus. Jensen, working in his private research institute in Copenhagen, discovered that there were two smuts of barley, "loose smut" and "covered smut." Jensen confirmed the suspicion that loose smut of oats did not affect barley or wheat, and indeed demonstrated that there were three fungi responsible for these problems: Loose smut of wheat (*Ustilago tritici*), Loose smut of barley (*Ustilago nuda*), and Loose smut of oats (*Ustilago Avenae*). Previously it had been considered that only one species of smut fungus (*Ustilago segetum*) was responsible. Jensen discovered that not only did the hot-water treatment kill the spawn of smut fungi within cereal seeds, but it also killed smut spores between the kernel and the husk, and on the surface of the outer seed coat as well. It not only prevented loose smut, for which no remedy had previously been found, but it was as effective as copper sulfate against stinking smut disease of wheat. He later demonstrated that all of the smut diseases of small grains, loose and covered, could be prevented by soaking the seed for half a day in warm water (68–86°F), and then plunging it for 5 min into hot water between the precise limits of 127–137°F, depending on the type of grain.

Jensen's hot-water treatment for cereal smut control received widespread adoption in Denmark, and even penetrated the conservative academic circles of the British Isles. Jensen's classical paper, "The Propagation and Prevention of Smuts in Oats and Barley," was published in the Royal Agricultural Society Journal in 1888. It was in the U.S.A. that the hot-water treatment was thoroughly investigated and then introduced into general agricultural use by Swingle in 1892. The contributions of Jens Jensen thus pioneered the use of thermal sterilization for the elimination of phytopathogens causing certain diseases of some economic crops.

Prior to World War II, inorganic mercuries such as mercuric chloride (corrosive sublimate) were widely used for seed treatment of vegetables. Following World War II the organic mercuries were developed and introduced as chemical seed protectants. Products such as Ceresan M, Ceresan L, Semesan, and Semesan Bel were widely used for vegetable seed treatment and for treatment of small grain seed.

The unfortunate incident of mercury poisoning of a family in New Mexico as a result of consuming mercury-treated wheat aroused a nationwide clamor to ban mercury chemicals from agricultural usage in the U.S.A.

The New Mexico mercury poisoning incident triggered nationwide concern about the use of chemicals in agriculture. Very few people were informed of the true facts concerning this unfortunate occurrence. The facts have been documented as follows. In August, 1969, a man drove approximately 100 miles to steal screenings from a plant using mercury to treat wheat seed. He fed these mercury-treated screenings to the family hogs. In September, 1969, one of these hogs became ill, so was butchered and fed to the man's family. Seven members of the nine-member family ate the pork. In October, the remaining 14 hogs also became sick. Twelve of them died and two hogs became blind. But the hogs were butchered and fed to the family. In December, 1969, three of the seven members of the family became ill—victims of mercury poisoning. In February, 1970, the news media reported this case of misuse of mercury-treated wheat screenings. The very next day, the USDA suspended the use of Panogen as a mercury-containing seed treatment. On July 15, 1970, the courts reinstated Panogen because they said the government ruling had been "arbitrary and capricious." Since that time, mercury has been found in undesirable quantities in fish, in lakes, and in streams. However, this is a different type of mercury and is not contamination from fungicide use. Many people considered the mercury in lakes and streams to be a pollutant from manufacturing plants. Of course, fish containing excessive mercury residues should not be used for food. It is interesting to note here that a scientist from the University of Chicago removed some preserved fish from a museum shelf where they had been for several decades and tested them for mercury content. They also had a high mercury content and were alive long before mercury fungicides were developed.

The ultimate result of the New Mexico mercury poisoning incident was the banning of all mercury fungicides for agricultural use in the U.S.A. The popularity of the pesticide issue has now reached the stage where almost everything that happens to fish, birds, deer, and sometimes people is blamed on pesticides. Even creatures as large as dinosaurs, mammoths, and other prehistoric animals failed in their struggle for survival. It has been estimated that 90% of all species of life that have existed at one time or another on this earth are now extinct.

Following World War II, when a shortage of mercury occurred, the introduction and acceptance of the first organic, non-mercurial seed protectant resulted. This was the product known as "chloranil," distributed under the trade name SPERGON in 1938, with tetrachloroparabenzoquinone as the active toxicant. During the 1930's Captan appeared as a competitor for thiram, and finally the oxathiins, distributed under the trade names Plantvax and Vitavax, were introduced as systemic seed treatments for smut diseases of small grains and corn.

WHY TREAT SEED WITH CHEMICALS?

The seed of any plant is one of the most fascinating and puzzling mysteries of biology. Within a tiny seed are all of the potentials for growth, development, and reproduction of a mature plant. From a small pinch of tobacco seed, small and light as a puff of dust, an acre of tobacco plants producing a ton of leaf for cigarettes and cigars can be grown. A single acorn, if it survives the vagaries of biological competition, can produce a mighty oak. Most plants have an infinite capacity for producing seeds in almost unlimited numbers. For example, a single plant of red clover only a few inches tall will produce 500 tiny copies of itself. Each weedy crabgrass, all too familiar to most home lawn admirers, produces 80,000 seeds. Pigweed rivals crabgrass by blowing and floating into farm fields and home lots with a million seeds per plant. It has been estimated that a single orchid can produce as many as 3,770,000 seeds. Plants produce seeds in such quantities because the mortality rate is high for infant plants.

Numerous plant diseases live over from year to year because spores of the causal pathogen are carried on the surface of the plant seed. When such contaminated seed is planted the seed-borne fungi and bacteria revive, start to grow, and become active parasites, destroying the seed in the soil or attacking the seedling as it struggles through the soil. Diseases of this type are referred to as "seedling blights." One purpose of chemical seed treatment is to inactivate surface-borne fungi and bacteria that may cause seed decay or seedling blights (Fig. 7.1).

Still other disease-inciting fungi may be borne within the seed, and as

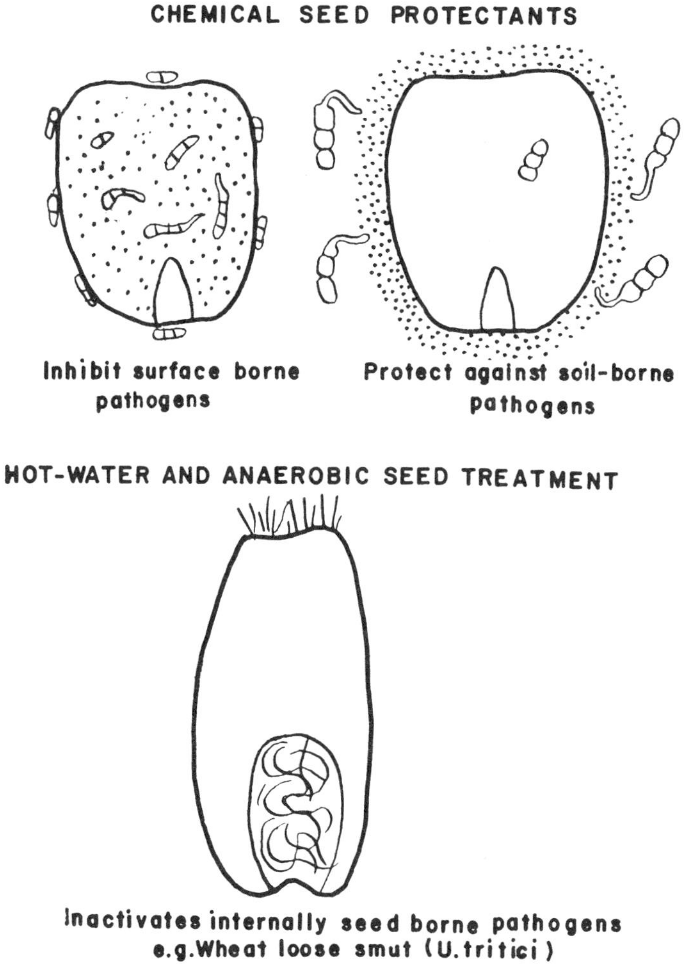

FIG. 7.1. THE PURPOSE OF CHEMICAL SEED PROTECTION

the seed sprouts the fungus grows up inside the developing plant, producing disease symptoms in the mature plant stage. Loose smut of wheat and barley are caused by fungi of this type. Seed treatment with hot-water or anaerobic cold-water treatment will effectively prevent development of internally seed-borne fungi. Finally, many microscopic molds and bacteria live in the soil and some of these are phytopathogenic. Chemical seed treatment is used to create a protective zone around the seed through which soil-borne microorganisms cannot penetrate, thus protecting the developing seedling against attack until it is established on a well developed root system, and has acquired the capability of outgrowing attacks by soil-borne pathogens.

TYPES OF SEED TREATMENT

There is no one seed treatment method or material that may be recommended as a universal preventative for all of the problems that can attack seeds and seedlings. Certain liquid seed treatments recommended for small grains would be ineffective and even harmful when used on some vegetable seeds. Since some of the seed treatment fungicides may be slightly toxic to animals and humans, care must be exercised when handling these products. Even though some seed treatment chemicals may not be toxic, most of them, like house dust, can be irritating to the eyes and nose. Chemically-treated seeds should NEVER be used as feed for livestock or food for human consumption, even though it may have been stored for months or years after chemical treatment. Containers that have been used for storing chemically-treated seed should be discarded or thoroughly cleaned before being used for other purposes.

Extreme care should be exercised to avoid inhaling dusts or fumes of seed treatment chemicals when applying them. For this reason treating of seeds with chemicals should be done either out-of-doors or in a well-ventilated room. Commercial seed-treating companies should provide an adequate air-exhaust system for seed treatment rooms, and workers exposed to chemicals for long periods of time should use an approved dust mask in which the filter can be changed frequently.

The precautions for the safe handling of seed treatment chemicals are clearly outlined on the can, bag, or container in which these products are packed. These instructions are required by federal law and are designed to protect the health and safety of the user.

Using more than the prescribed dosage for the specific seed may cause injury to the seed. Using less than the prescribed dosage may result in disappointing control of the diseases for which they are recommended. Growers and home gardeners who use seed treatment chemicals with the prescribed precautions need have no fear concerning their hazards. Seed treatment with chemicals is a useful tool in our arsenal of defense against losses caused by some plant diseases.

The different types of seed treatment for plant disease control may be classified as follows:

1. Hot-water treatment for small grains
2. Cold-water (anaerobic) treatment for cereals
3. Hot-water treatment for vegetable seeds
4. Organic chemical dust and wettable powder treatments (see Table 7.1)

The Hot-Water Treatment for Small Grains

The hot-water treatment is the only method known at the present time that will control the loose smut diseases of wheat and barley (*Ustilago tritici* and *Ustilago nuda*). Because both of these pathogens are carried internally within the seed, other chemical seed treatments will not prevent the diseases they cause. The hot-water treatment is tedious, laborious, exacting, messy, and time-consuming. For this reason, it is suggested that only sufficient seed to plant a small acreage be treated from which nearly smut-free seed may be harvested for general seeding the following year. The hot-water method of seed treatment is no task for the amateur or individual farmer who rarely has adequate equipment for the process. If a live-steam supply is available, better temperature control can be maintained and a larger amount of seed treated at one time. However, the technical requirements for a satisfactory job of treating seed by the hot-water method are the same whether hot water or steam is used as the source of heat. The procedure for the hot-water method of seed treatment is as follows:

a. Place the seed to be treated in half-bushel lots in burlap sacks. Tie the sacks loosely at the top because grain will swell during the treatment.
b. Pre-soak for 5–6 hours in water at 60–70°F (15–20°C).
c. Dip for 5 min in water at a temperature of 120°F (49°C) to pre-heat the seed.
d. *For wheat:* Immerse the seed in hot water at 129°F (54°C) for 10 min.
   For barley: Immerse the seed in hot water at 126°F (52°C) for 13 min.
   Agitate the grain from end to end of the sack during treatment in order to sustain uniform temperature throughout the sample being treated. It is extremely important to maintain the specified temperatures exactly during the treatment.

TABLE 7.1. REGISTERED SEED TREATMENT FUNGICIDES FOR VEGETABLES

There are two types of seed treatment for vegetable crops:

1. Eradicative treatments for the elimination of seed-borne pathogens
2. Protective seed treatments are applied to the seed surface to protect the seed from seed decay and from invasion of seedlings by soil-borne pathogens

Eradicative seed treatment: Since the use of mercury-containing fungicides such as mercuric chloride (bichloride of mercury or corrosive sublimate) is now prohibited by federal law, the hot-water treatment is the only eradicative treatment that may be legally used on vegetable crops.

The hot-water treatment, when used correctly, will eradicate most of the seed-borne pathogens of vegetables. It is most widely used for the elimination of bacterial seed pathogens and is suggested for use on the following vegetable seeds: cabbage, carrot, cucumber, eggplant, lettuce, pepper, radish, spinach, tomato and turnip.

When used improperly, the hot-water treatment can cause severe seed injury. Seeds of cucurbits are especially susceptible to injury from this type of treatment. The following procedure is suggested for effective use of the hot-water treatment:

Fill a loosely woven cotton bag not more than half-way with the seed to be treated and pre-warm for 10 min in 38°C (100°F) water. Next, immediately place the pre-warmed seed in a water bath in which it is possible to hold the temperature exactly where recommended for specific vegetables. It is most important to maintain the water at the exact recommended temperature for the treatment interval. After the hot-water treatment is completed, dip the bag of treated seed in cold water to stop the heating action, and spread the seed out to dry on a suitable smooth surface. Hot-water treated seed should always be overcoated with an approved fungicide before planting. Old seed may be severely injured by the hot-water treatment, and for this reason, when seed more than one-year-old is hot-water treated a seed germination test should be made before planting.

The following temperatures and treatment time intervals are suggested for the hot-water treatment of vegetables. It is most important to follow these suggested temperatures exactly in order to avoid seed injury.

Crop	Hot-Water Temperature	Treatment Time Interval
Broccoli, cauliflower, Chinese cabbage, cucumber, carrot, collard, kale, kohlrabi, rape, rutabaga, turnip	122°F (50°C)	20 min
Brussels sprouts, cabbage, eggplant, spinach, tomato	122°F (50°C)	25 min
Mustard, cress, radish	122°F (50°C)	15 min

Pepper	125°F (51.7°C)	30 min
Lettuce, celery, celeriac	118°F (47.7°C)	30 min

Vegetable seeds, like other plant seeds, are subject to attack by a number of seed-borne and soil-borne pathogens which may initiate seed decay and seedling blights, if weather conditions at planting time favor plant disease development.

Several fungicides are registered and approved for use on vegetable seeds to protect them against seed decay and seedling blights. Commercially packaged vegetable seeds are frequently pre-treated with fungicides prior to distribution. Pink-colored lima bean, cucumber, beet, and pea seeds indicate they have been protected with a fungicide seed treatment.

For the convenience of the home gardener or the commercial vegetable grower, suggestions for the use of seed protectants registered for use on vegetable seeds are summarized below.

Crop	Suggested Fungicide	Dosage: Per 100 lb of seed	Dosage: Per 1 lb of seed
Beans (snap and lima)	Thiram 75% WP	1.5 oz	0.5 tsp
	or Dexon 70% WP or	1.0 oz	0.25 tsp
	Captan 50% WP	1.5 oz	2.0 tsp
Beets	Dexon 50% WP	3.0 oz	2.0 tsp
	or Thiram 50% WP	3.0-4.0 oz	1.5 tsp
	or Captan 50% WP	3.5 oz	2.0 tsp
Cabbage, cauliflower, broccoli and Brussels sprouts	After hot water treatment treat with		
	Thiram 75% WP	3.0 oz	1.0 tsp
	or Captan 50% WP	1.8 oz	0.75 tsp
Carrots	Thiram 50% WP	3.0-4.0 oz	1.0 tsp
	Captan 50% WP	4.0 oz	1.5 tsp
Celery and parsnips	Hot-water treatment	Or use seed at least 2 years old	
Cucumber, melon, squash, and pumpkin	Thiram 75% WP	3.75 tsp/5 lb	0.5 tsp
	or Captan 50% WP	2.5 tsp/5 lb	
	or Dexon 70% WP	1.25 tsp/5 lb	

TABLE 7.1 *(Continued)*

Crop	Suggested Fungicide	Dosage: Per 100 lb of seed	Dosage: Per 1 lb of seed
Eggplant	After hot-water treatment treat with Thiram 75% WP	4.0 oz	0.67 tsp
Lettuce, endive and escarole	Thiram 75% WP	3.0 oz	0.67 tsp
Onions	Use seed pelleted with Captan 50% WP or Thiram 75% WP	10.67 oz/ 10 lb seed	Where smut is a problem
	Captan 50% WP or Thiram 75% WP	4.0 oz	Where smut is not a problem
Peas	Thiram 75% WP	2.0 oz	0.25 tsp
	or Dexon 70% WP	0.7 oz	
	or Ortho 65 (Captan + graphite) at manufacturer's direction		
Pepper	Thiram 75% WP	5.5 oz	1.0 tsp
	or Captan 50% WP	4.5 oz	1.0 tsp
Potato	Dip or dust cut or whole tubers in Captan 50% WP 0.5-1.5 lb in 100 gal. water as a dip or Captan 50% WP 0.75-1.5 oz/100 lb tubers or Maneb 70% WP 1.0 lb/100 gal. water as a dip or Manzate 200 2.0 lb/100 gal. water as a dip		
Spinach	Thiram 50% WP	3.0–4.0 oz	1.5 tsp
Sweet Corn	Thiram 50% WP	2.5–4.0 oz	1.75 tsp

	or Captan 50% WP	0.5-2.0 oz	0.75 tsp
	or Dexon 70% WP	0.28-0.56 oz	0.75 tsp
Sweet Potato	Dip roots and sprouts for 5 min in Thiram 50% WP 0.75 lb in 7.5 gal. water. Plant immediately		
Tomato	Treat with hot-water for bacterial diseases. After seed dries treat with Thiram 50% WP	2.5-4.0 oz	1.25 tsp

tsp = teaspoonful.

e. Spread the treated seed out to dry on a clean, tight floor.
f. After the seed has dried treat with a recommended organic non-mercurial fungicide (see Table 7.1).
g. The hot-water treatment may lower germination. For this reason the seed germination should be checked before planting. Treatment may also delay seed germination and this should be taken into consideration when running seed germination tests.

The Cold Water (Anaerobic) Seed Treatment Method

This method is less injurious than the hot-water method and is suggested where adequate facilities for the hot-water treatment are not available. The cold-water treatment is suggested for the control of loose smut of wheat and barley where the causal fungus occurs within the seed rather than on the seed surface. The procedure is as follows:

a. Place the seed to be treated in half-bushel lots in burlap sacks or other types of loose mesh bags. Tie loosely at the top since the grain will swell during treatment.
b. Pre-soak in water at room temperature (60–70°F (15–20°C)) for 4 hours.
c. Drip-drain for 30 min and store seed in air-tight containers (such as 50-gal. metal drums with tight-fitting, clamp-style lids) for 70 hours at room temperature, or store at an average temperature of 68°F (20°C) for 80 hours, 75°F (24°C) for 60 hours, 80°F (27°C) for 40 hours, or 90°F (32°C) for 30 hours.
d. After the storage period drain off water and spread the seed out to dry on a clean, tight floor. Dry quickly to a normal seed moisture of 10–18% and do not permit seed to heat above 100°F (38°C) during drying.
e. After drying treat the seed with one of the recommended organic non-mercurial seed protectants (see Table 7.1).

The Hot Water Seed Treatment for Vegetables

The hot-water treatment for vegetables is for the control of diseases such as Blackleg (*Phoma lingam*) and Black Rot (*Xanthomonas campestris*) of cabbage and other crucifers, celery early and late blights (*Cercospora apii* and *Septoria apii*), eggplant anthracnose (*Colletotrichum* sp), and tomato anthracnose (*Colletotrichum phomoides*). The procedure is as follows:

a. Place the seed to be treated in a suitable loose mesh cotton bag or other container. Do not fill more than half-full.

b. Soak the seed for the specified time at the prescribed temperature. It is most important to maintain the recommended temperatures accurately for the duration of the treatment period. An accurate laboratory or dairy type thermometer should be used for adjusting temperatures. A temperature of only 2–3° above that specified may cause serious injury to seed germination or the treatment may be ineffective. The prescribed temperature exposures for different vegetables are given in Table 7.1. After completion of the hot-water treatment the seeds should be thoroughly dried at 70–75°F (21–24°C) and then treated with a recommended organic non-mercurial seed protectant (see Table 7.1).

METHODS OF APPLYING SEED PROTECTANTS

Seed protectants may be applied to seed in small lots by the farmer or homeowner using relatively simple homemade treaters, or they may be applied to large seed lots by commercial seedsmen using expensive and specialized seed-treating equipment.

The simplest of the homemade seed treaters is the *oil-drum or rotary seed treater* (Fig. 7.2). The rotary treater may be made from a 30-gal. oil drum and readily obtainable accessories. One-half of one of the heads of a steel oil drum is first cut out with a chisel, hacksaw, or welding torch.

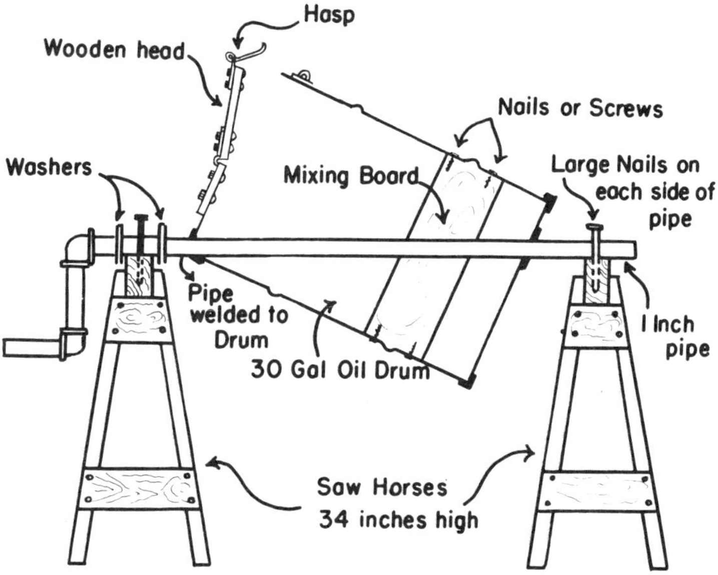

FIG. 7.2. THE OIL-DRUM OR ROTARY SEED TREATER

TABLE 7.2. REGISTERED SEED TREATMENT FUNGICIDES FOR WHEAT, BARLEY, OATS AND RYE

(Prepared by Einar W. Palm, Department of Plant Pathology, University of Missouri, Columbia)

Fungicide	Trade Names and Manufacturers	Formulation	Dosage and Comments
CAPTAN	Captan 25 (25% Captan), Stauffer; Corn States	Dust or planter box	3 oz/bushel. Protects against seed-borne and soil-borne pathogens on wheat. Not effective for the control of stinking smut or loose smut of wheat
	Captan 80	Slurry	2-3 oz/bushel on wheat, barley, and oats. Not registered for rye. Protects against several seed-borne and soil-borne pathogens. Not effective for smut control
	Evershield Seed Protectant CM (29.52% Captan, 0.34% Malathion), Gustafson Mfg., Inc.	Slurry	1 gal. CM protectant concentrate in 5 gal. cold water, 320 cc/cwt for barley; 420 cc/cwt for oats; 270 cc/cwt for wheat; 370 cc/cwt for rye applied with a slurry treater. Protects against several seed- and soil-borne pathogens. Not effective against smuts
	Ortho Seed Protectant (45% Captan, 40% HCB), Chevron	Dust, slurry, planter box	1 oz/bushel for wheat, oats, barley and rye. Some stinking smut and covered smut control. Protects against many seed- and soil-borne pathogens. Not effective for the control of loose smut of wheat and barley
CARBOXIN	Vitavax (75% Carboxin), Uniroyal	Dust or slurry	4 oz/cwt for wheat and barley only. Controls all smuts including loose smut. Use as an overcoat. Do not use crop from treated seed for hay, straw or grain as food or feed. Do not graze
CARBOXIN COMBINATIONS	Vitavax-200 (17% Carboxin, 17.0% Thiram), Uniroyal, Ortho, Gustafson	Dust, slurry or liquid treatment	3-4 oz/cwt on wheat and barley only. Protects against smuts and seedling blights
MANEB	Manzate (80% maneb), DuPont	Dust or planter box	2 oz/bushel for wheat, barley and rye (not registered for oats). Controls covered smut or bunt of wheat,

			barley and rye. Not effective against loose smut of wheat and barley. Use Vitavax overcoat treatment
	DB-Green (50% maneb, 18.75% lindane), Agsco	Dust	2 oz/bushel on wheat, barley and rye. Protects against several seed and seedling pathogens. Effective against stinking smut of wheat and covered smut of barley. Will not control loose smut of wheat and barley. Some insect protection 1 oz/bushel for wheat. Not effective agaist loose smut of wheat and barley. Same control as Granox N-M
MANCOZEB	Dithane M-45 Seed Protectant (80% maneb plus zinc ion), Rohm & Haas or Manzate 200 (80% maneb plus zinc ion), DuPont	Dust or slurry	1⅓ oz/bushel as a slurry treatment for wheat, barley, oats and rye. Protects against several seed and soil-borne pathogens. Effective against stinking smut and covered smuts. Not effective against loose smut of wheat and barley
TERRACLOR + TERRAZOLE	Terra-Coat L 205 (23.2% PCNB, 518% Terrazole), Olin	Liquid	2 oz/bushel for wheat (not registered for oats and rye). Controls covered smut, stinking smut and seedling blights. Does not control loose smut of wheat and barley
	Terra-Coat SD 205 (20% PCNB, 5% Terrazole), Olin	Dust	3 oz/bushel for wheat, barley and oats. Not registered for rye. Controls covered smuts and stinking smut of wheat and seedling blights. Not effective against loose smut of wheat and barley
THIRAM	Arasan 50-Red (50% Thiram),	Dust, slurry and	2 oz/bushel for wheat, barley, and rye. Not registered for oats. Controls covered smut and stinking smut

TABLE 7.2 *(Continued)*

Fungicide	Trade Names and Manufacturers	Formulation	Dosage and Comments
	DuPont	planter box	of wheat. Overcoat treatment of Vitavax is suggested for loose smut control
	Arasan 70-S (70% Thiram, 2% methoxychlor), DuPont	Slurry	1⅓ oz/bushel for wheat, barley and rye. Not registered for oats. Provides additional insect protection during pre-planting storage (rice weevil, confused flour beetle, grain moth and Indiana meal moth)

Seed treatment of small grains with fungicides is generally accepted to reduce losses caused by seed decay, seedling blights, cereal smuts, and several other seed-borne and soil-borne diseases such as Fusarium blight or scab, Septoria leaf blotch and glume blotch, Helminthosporium spot blotch and net blotch.

The information summarized in the foregoing table is merely a compilation of fungicide choices for seed treatment of small grains and is not intended as a recommendation.

Small grains treated with fungicides must not be used for human consumption or as animal feed. Federal law requires chemically treated small grains to be labelled: THIS SEED TREATED WITH A FUNGICIDE (name of product). Do not use for food, feed or oil purposes.

TABLE 7.3. REGISTERED SEED TREATMENT FUNGICIDES FOR SOYBEANS

(Prepared by Einar W. Palm, Department of Plant Pathology, University of Missouri, Columbia)

Seed treatment of soybeans with a fungicide before planting has been demonstrated to improve germination and seedling vigor by reducing seed decay, damping-off and seedling blights. Soybean seeds often carry fungus spores of Aspergillus, Penicillium, Rhizopus, and other molds that may affect germinating seed. In addition, several fungi that cause soybean diseases—such as pod and stem blight, anthracnose, purple seed stain and downy mildew—may severely affect the performance of soybean seed. In some years harvesting conditions result in cracks in seed coats that may open the door for invasion by seed-borne and soil-borne pathogens, resulting in seed rots and seedling diseases. There is some disagreement among plant pathologists relative to the value of chemical seed treatment of soybeans. However, in recent years there has been increasing acceptance of chemical seed treatment of soybeans by commercial growers who have enjoyed profitable returns from this practice. If less than 80% seed germination is due to disease-inciting pathogens, seed treatment will usually improve field performance and subsequent yields. If poor germination is the result of mechanical injury to the embryo, or other physiological disorders, chemical seed treatment of soybeans may not be profitable but it may reduce loss of stand caused by soil-borne pathogens. If seed germination is more than 90%, seed treatment may or may not result in sufficient benefits to justify the expense, depending on environmental conditions at planting time. The following summary of chemical seed treatments for soybeans is intended as a guide and not a recommendation for commercial producers of soybeans.

Fungicide	Trade Names and Manufacturers	Formulation	Dosage and Comments
CAPTAN	Captan 25 Seed Protectant, Stauffer	Dust	4.0 oz/bushel or 7.0 oz/cwt as a planter box treatment
	Captan-Moly (25% Captan, 4.75% molybdenum), Stauffer	Dust	4 oz/bushel applied as a planter box treatment. Soybeans should be planted at the rate of at least 1.0 bushel per acre for optimum molybdenum capability for nodule bacteria and is most effective below pH 6.0
	Captan 80 Seed Protectant (80% Captan), Stauffer	Slurry	1.0 oz/bushel. Pre-mix recommended amount with sufficient water to make a thin paste. Stir paste into the amount of water required for the slurry treater to be used.
	Orthocide 75 Concentrate (75% Captan), Chevron	Dust or slurry	2 oz/cwt as planter box treatment, or 1⅓ oz/cwt applied in a slurry treater
	Evershield CM Seed Protectant (29.52% Captan, 0.34% Malathion), Gustafson	Slurry	1 gal. CM Protectant Concentrate in 5 gal. cold water plus 7 oz Evershield dye. 240 cc of diluted product per 1 cwt of seed. Provides some insect protection
CHLORONEB	Demosan 65W (65% Chloroneb), DuPont	Slurry	4 oz/cwt used as an overcoat with Captan- or Thiram-treated seed, or as a combination treatment using a slurry treater

TABLE 7.3 *(Continued)*

Fungicide	Trade Names and Manufacturers	Formulation	Dosage and Comments
CARBOXIN-THIRAM	Vitavax-200 Flowable Fungicide (17% Carboxin + 17% Thiram), Uniroyal-Gustafson	Slurry	4 oz/cwt used as overcoat with Captan- or Thiram-treated seed, or as combination treatment in slurry treaters
TERRACLOR	Terra-coat SD 205 (20.0% PCNB, 5.0% Terrazole), Olin	Dust or slurry	2-4 oz/bushel used as a slurry or planter box treatment
	Terra-coat L 205 (23.2% PCNB, 5.8% Terrazole), Olin	Liquid	2-4 oz/bushel. Use directly from container or it may be mixed with water for use in liquid seed treating equipment
THIRAM	Arasan 50 Red (50% Thiram), DuPont	Dust	2 oz/cwt applied in a commercial dust treater, home-made rotary drum or planter box treatment
	Arasan 70-S (70% Thiram, 2% methoxychlor), DuPont	Slurry	1⅓ oz/cwt applied with a commercial slurry treater. Do not use in commercial dust treaters
	Arasan 42-S (42% Thiram + special sticker), DuPont	Liquid	2 oz/cwt using slurry treater

Seed treatment fungicides for soybeans are formulated for use as dry treatment of seeds (dusts) to be applied by commercial dust treaters, by homemade rotary seed treating drums, or by mixing directly with the seed in the planter box. Some formulations are wettable powders which may be used in slurry treaters. Only a few formulations are liquids.

Use of the rotary drum for seed treatment: Fill the unit approximately half-full of seed; add the recommended dosage of fungicide, and rotate slowly until color is uniformly imparted to all seed. Handle seed carefully to prevent cracking.

Planter box seed treatment: Fill the planter box half-full and sprinkle half of the recommended seed treatment fungicide over the surface of the seed. Mix by stirring with a wooden paddle until a uniform color is imparted to the seed. Add the remaining seed to fill the planter box, then sprinkle the remaining half of the necessary fungicide, and mix uniformly with the wooden paddle. To prevent plugging of the planter plate, omit seed treatment chemical every fourth filling. Planter box treatments are somewhat ineffective for air blast planters. Dust-free formulations such as Evershield are more satisfactory for air-blast planters.

WARNING: ALL SOYBEAN SEED TREATED WITH A FUNGICIDE MUST BE LABELLED "THIS SEED TREATED WITH (E.G., CAPTAN) FUNGICIDE. FEDERAL LAW PROHIBITS THE USE OF CHEMICALLY TREATED SEED FOR FOOD, FEED, OR OIL EXTRACTION"

The door or lid of the treater is made from a strip of 1-in. board, 6 in. wide, and approximately 17 in. long. This is bolted to the remaining half of the steel drum head so that 1 in. of the steel head extends beyond the board towards the opening. A semicircular board, 1 in. thick and of the same length as the diameter of the drum, and 1 in. wider than one-half the diameter of the drum (approximately 10 in.) is then cut to fit snugly into the open half of the drum head. A pair of 8-in. strap hinges are attached with carriage bolts to the semicircular wooden head so that it will serve as a lid for the treater. The wooden head or lid must be made dust-tight by nailing weather stripping or heavy felt around the edge. A hinged hasp is used to hold the door or lid shut and is attached so as to force it tightly against the edge of the steel drum when closed. A mixing board, or baffle, made from a piece of smoothly planed 1-in. board, 8 in. wide and 18 in. long, is then placed across the inside of the drum. The baffle board should be attached approximately two-thirds of the way back from the opening and at right angles to the straight side of the door. The completed seed-treating drum is then attached to a pair of wooden saw-horses using 1-in. pipe for an axle and crank. To use this homemade device it should be filled no more than one-half full of seed. The necessary amount of chemical seed protectant is then added, and the barrel slowly rotated for at least 25 revolutions.

Another simple homemade seed treater is the *gravity seed treater* (Fig. 7.3). This is constructed from a suitable steel drum equipped with a system of cones or baffles as illustrated. In this equipment the seed and the chemical seed protectant are added simultaneously and allowed to flow down by gravity over the baffle cones, which accomplishes the treating process. The gravity-type treater can handle 30–60 bushels of small grain seed per hour.

The Minnesota seed treater (Fig. 7.4) is another popular gravity treater constructed of wood to the dimensions given with 5 baffles accurately located, as indicated. To build this type of seed treater, first cut all of the parts shown using 1-in. × 10-in. boards planed on both sides and edges. The mixing chute should be nailed together as shown, leaving the placing of the baffle boards until last. The sides "a" and "c" and the door must overlap the front and back "b" and "d." The upper edges of the baffles are then bevelled to 45° to fit snugly inside the chute. A 2-in. hole is bored in each baffle, 2 in. from the top of the longer side and ¾ in. from the right-hand edge as one faces the longer side. Baffle No. 1 is placed with its bevelled edge against "a," 17½ in. from the top of the chute, with the hole under the opening that will be made by raising the trough. Baffle No. 2 is placed against side "b," 24 in. from the top of the chute; Baffle No. 3 against side "c," 29½ in. from the top; Baffle No. 4 against side "d," 36 in. from the top, and Baffle No. 5 against side "a," 41½ in. from the top. When the mixing chute has been completed, study the way the proportioning trough will fit into it and the manner in which the two will fit together. Nail up the trough, being careful to overlap the boards as shown. The two long side boards overlap the bottom board, and the two short bevelled side boards overlap the long ones and rest on top of the bottom boards. Finally, the trough and chute are put together, push-

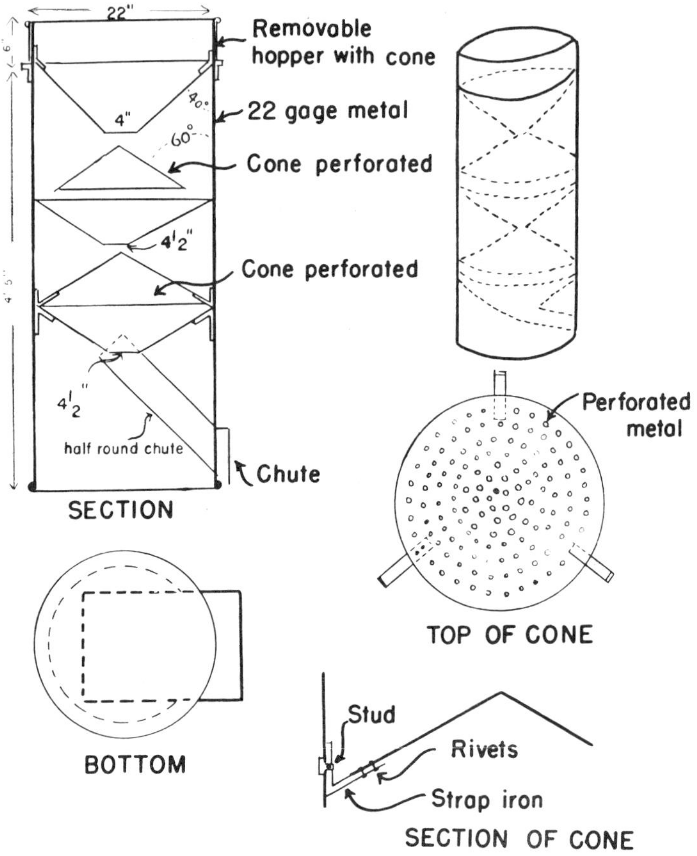

FIG. 7.3. THE GRAVITY SEED TREATER

ing the trough into the front of the chute so that no opening is left in the chute unless the trough is raised. The trough holds 2½ bushels and is designed to treat 2 bushels of grain at a time. To operate the Minnesota seed treater a bushel of grain is distributed at an even depth along the mixing chute. The seed treatment compound is then sprinkled over the surface of the seed and a second bushel of grain is then distributed in the trough. The mixing chute is then tipped forward, allowing the grain to pass through the baffle arrangement and in so doing becomes well mixed with the seed treatment chemical.

The slurry seed treater was introduced by the DuPont Company at the end of World War II. With this equipment a slurry composed of the seed treatment chemical and water is metered by a simple mechanism consisting of a slurry cup and a seed dump pan. The slurry cup introduces a

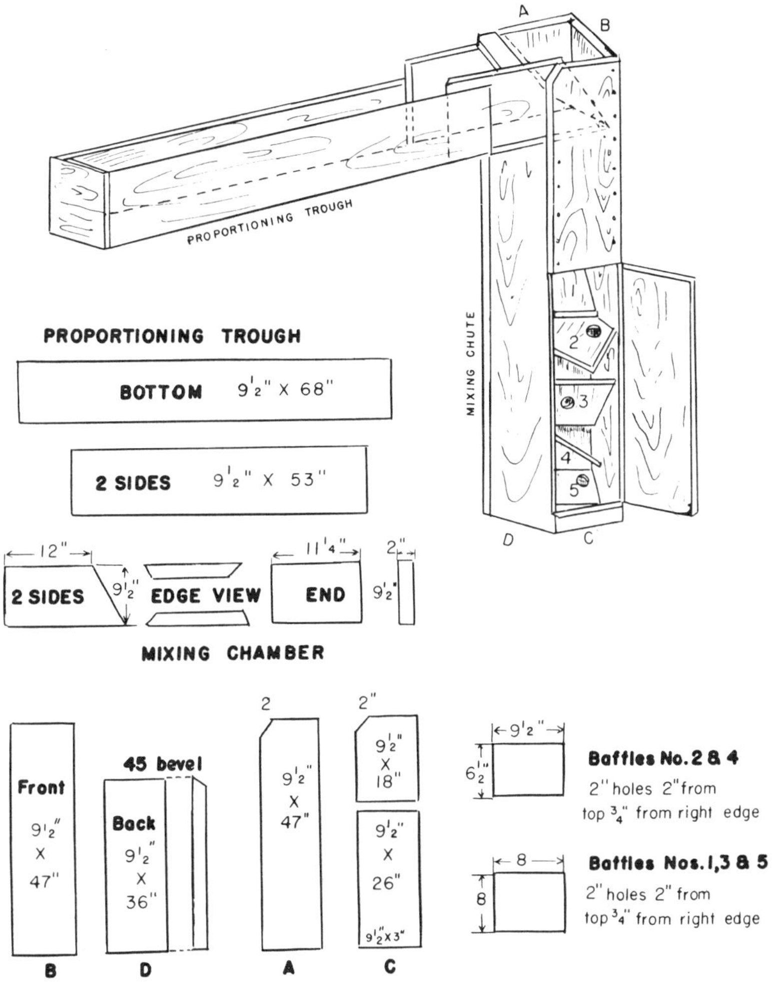

FIG. 7.4. THE MINNESOTA SEED TREATER

given amount of slurry with each dump of seed into a mixing chamber where they are thoroughly mixed. The slurry treater eliminates objectionable dusts of seed treatment chemicals. All types of seeds may be treated with this type of equipment which has a capacity of up to 600 bushels per hour. The slurry seed treater is illustrated in Fig. 7.5.

In the slurry treater the seed is metered in specified amounts by weight, varying for different types of seeds, by means of a tilting seed pocket, and flows into the treating hopper from an overhead bin. When the correct weight of seed has flowed into the seed pocket, it tilts and dumps

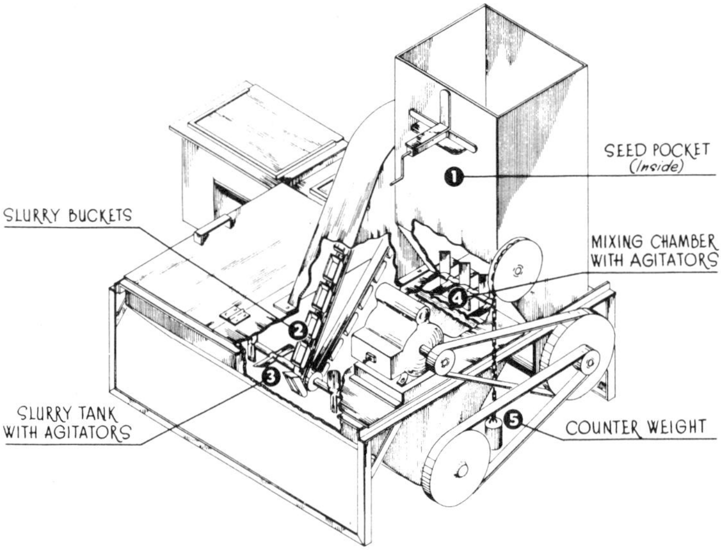

FIG. 7.5. THE SLURRY SEED TREATER

the load into the mixing chamber. A counterweight returns the seed pocket to its original position. The two movements, downward to unload and upward to reload, are activated by means of a shaft attached to the back edge of the pocket to an endless sprocket chain equipped with small slurry cups. The chain runs over two sprocket wheels. One of these, equipped with 15 teeth, is idle on the shaft, while the other, provided with 9 teeth, is located on a short shaft on the bottom of the tank containing the slurry mixture. Each time the seed pocket dumps a load of seed, the shaft turns the 15-tooth sprocket by means of a rachet wheel through an arc of 3 teeth. Since every third link on the sprocket chain carries one slurry cup, each dumping of the seed pocket advances one slurry cup and dumps its contents into the mixing chamber with the seed. Each slurry cup holds sufficient slurry to treat the amount of seed metered by the tilting seed pocket. The newer types of slurry treaters are equipped with a single chain of cups with parts and rubber plugs for 15 cc, 25 cc, and 46 cc quantities. With all plugs removed the cup delivers 15 cc of slurry; with the bottom port plugged it delivers 25 cc of slurry; and with two plugs inserted it delivers 46 cc of slurry mixture.

The mixing chamber of the slurry treater is fitted with an auger type agitator that mixes and propels the seed to the bagging end of the mixing chamber. The auger may be of several types—curved paddles, curved rods, or nylon brushes. The nylon brush auger is advisable for thin-coated seeds and seeds that have a tendency for mechanical injury. The construction of the auger is important since the slower it moves the seed through the mixing chamber, the more uniform the distribution of the slurry mixture on the seed being treated. The slurry treater is one of the most precise and accurate devices for applying wettable formulations of seed protectants to seed.

Direct ready-mix or automatic seed treaters are used in larger seed-treating plants, such as grain elevators, and were developed to handle liquid formulations of fungicide seed protectants. *Planter box seed treatment* formulations have made possible seed treatment on the farm by simply adding the seed treatment chemical to seed in the planter box, and eliminates the need for special seed treatment equipment.

8

Antibiotics for Plant Disease Control

An antibiotic has been defined as "a chemical substance produced by an organism which possesses inhibitory or germicidal properties against other microorganisms." The word "antibiotic" is the converse of "symbiotic," and literally means "against life." Gottlieb and Shaw (1970) have defined antibiotics as "organic substances that are produced by microbes and are deleterious at low concentrations to the growth and metabolic activity of other microorganisms. The term also includes organic compounds produced by higher plants." The use of antibiotics for the treatment of human and animal diseases has been a major breakthrough in medical science and in veterinary medicine for the treatment of maladies of bacterial etiology.

The discovery of penicillin is attributed to the Scottish bacteriologist Sir Alexander Fleming. In 1929 he observed that a culture of *Staphylococcus aureus* failed to develop satisfactorily when it was contaminated with *Penicillium* sp, the common blue mold fungus. *Penicillin* is a common laboratory contaminant and saprophyte on stored fruits and vegetables. The research of Sir Alexander Fleming later demonstrated the antibiotic properties of this fungus by-product against a wide range of bacterial organisms. He named the by-product "penicillin," and launched the era of antibiotics that was destined to become a major weapon in the treatment of bacterial diseases of man and animals. The discovery of penicillin, produced from the mold *Penicillium notatum*, opened the door for the exciting and widely used "miracle drugs" that are now routine treatment for many human and animal diseases.

Since the introduction of penicillin, several thousand compounds with varying degrees of antibiotic activity have been developed. In addition to their use for the treatment of human and animal diseases, some have been used for prevention of spoilage in processed foods, vegetables, and dairy products.

Ten years elapsed after Fleming's discovery of penicillin before extensive clinical tests were conducted and research on methods for volume production was initiated by Florey (1949) and his associates. The 1945 Nobel prize in Physiology and Medicine was shared between Sir Alexander Fleming and the Australian pathologist Sir Howard Walter.

According to Torgeson (1967) the word "antibiosis" was first used by Vuillemin in 1889 and was adopted ten years later by the British plant pathologist Marshall Ward to describe certain inhibition phenomena in the development of fungi causing plant diseases.

The first attempt to control a plant disease by the use of an antibiotic was the work of Potter (1908). In 1908 he reported that the concentrated liquid from a nutrient culture of *Pseudomonas destructans*, the pathogen responsible for the bacterial rot disease of turnips, would inhibit the organism itself when it invaded turnip tissue. He also reported that by-products of *Penicillium italicum* would prevent secondary organisms from invading the skin of stored oranges. In his article published in 1908 Potter stated, "Many other parasites may be checked, if not effectively prevented from further progress by the adoption of this principle (viz: the use of antibiotic fungus by-products). In 1926 Guthrie B. Sanford, plant pathologist with the Science Service, Canada Department of Agriculture, at the University of Alberta, Edmonton, Canada, concluded that control of the potato scab disease (*Actinomyces scabies*) by green manure crops, was "due to the antibiotic qualities of certain predominant soil microorganisms."

STREPTOMYCIN

Streptomycin, a by-product of the actinomycete *Streptomyces griseus*, was discovered in 1944 by Schatz, Bugie and Waksman. In 1946 eradication of *Xanthomonas pruni* (bacterial blight of peaches) by soaking peach budsticks in solutions of streptomycin was reported. Some degree of success in eradicating crown gall (*Agrobacterium tumefaciens*) on *Prunus* sp. by painting the galls with strong solutions of streptomycin was also reported. Murneek first reported effective control of fireblight of apples and pears with sprays of streptomycin sulfate in 1952. Since this early report many papers have been published supporting the observations of Murneek.

Streptomycin is available for plant disease control as streptomycin sulfate and streptomycin nitrate. It is an odorless, white, somewhat hygroscopic powder, readily soluble in water, but insoluble in most organic solvents. It is available for agricultural use under the following trade names: Agrimycin. 17, Agri-Strep, Chemform, Gerox, Hokko-Mycin, Phytomycin, and Rimocidin.

The basic producers of streptomycin in the U.S.A. are Charles Pfizer and Co., Food Machinery Corporation, Agricultural Chemical Division, and Merck Chemical Company.

Since the dosages for streptomycin products are given in parts per million, a conversion table for parts per million to standard measures is given in Table 8.1.

The control of plant diseases by the use of streptomycin is summarized in Table 8.2.

CYCLOHEXIMIDE

Cycloheximide was first reported by Whiffen as an antifungal antibiotic in 1948, and its agricultural uses were reported by Ford, Klomparens, and Hamner in 1958. Cycloheximide is a by-product of streptomycin-producing strains of *Streptomyces griseus* and is a representative of a group of antibiotics classified as glutarimides. Its antifungal properties were reported in 1948 by Felber and Hamner with the observation that 1.0–5.0 ppm would control powdery mildew on bean plants (*Erysiphe polygoni*). Its spectrum of antibiotic activity was further investigated in 1950 by Whiffen who reported its remarkable specificity of activity against microorganisms. He reported inhibition of growth of certain yeasts at concentrations of 0.1–1.0 ppm, while at 1000 ppm it failed to inhibit other yeast species and was non-biocidal to the 12 bacterial species included in the test.

In 1949 Peterson and Cation (1950), working at Michigan State University, reported remarkable control of cherry leafspot disease (*Coccomyces hiemalis*) with spray applications of 1–10 ppm of cycloheximide. This observation was confirmed by other investigators, and cycloheximide was introduced commercially in 1954 for use on non-bearing trees for the control of cherry leafspot disease. It was approved for this purpose by the U.S. Department of Agriculture and the Food and Drug Administration in 1955.

The use of cycloheximide for the control of turf diseases was initiated by Hamner and Vaughan in 1949 at Michigan State University. A solution of cycloheximide at a concentration of 300 ppm gave excellent control of melting-out disease of bluegrass (*Helminthosporium sativum*). Later, in New Jersey and Michigan, Davis, Engel, and Silber (1951) and Vaughn (1951) reported the effectiveness of cycloheximide for the control of large brown patch (*Rhizoctonia solani*) and dollarspot disease of bentgrass (*Sclerotinia homeocarpa*), respectively, and reconfirmed its value for the control of melting-out disease of turfgrasses. Since its introduction around 1950, cycloheximide, under the trade name Acti-Dione, has been the standard commercial treatment for the control of turfgrass diseases.

TABLE 8.1. CONVERSION TABLE FOR STREPTOMYCIN PRODUCTS

The dosages of streptomycin sulfate are given in parts per million (ppm). To convert parts per million to standard measures, use conversion factors below

Concentration Desired (ppm)	Quantity of 15–17% Streptomycin Sulfate Required				
	1 gal. (3.75 liters)	10 gal. (37.5 liters)	50 gal. (187.5 liters)	100 gal. (375.0 liters)	500 gal. (1875 liters)
50	0.5 tsp	0.5 oz (11.34)	2.0 oz (56.7)	4.0 oz (113.4)	1.25 lb (567.0)
60	0.6 tsp	0.6 oz (13.61)	2.4 oz (68.04)	4.8 oz (136.08)	1.5 lb (680.4)
100	1.0 tsp	1.0 oz (28.35)	4.0 oz (113.4)	8.0 oz (226.8)	2.5 lb (1134.0)
200	2. tsp	2.0 oz (56.7)	8.0 oz (226.8)	1.0 lb (453.6)	5.0 lb (2268.0)

tsp = teaspoonful.
Numbers in parentheses = grams.

TABLE 8.2. THE USES OF STREPTOMYCIN FOR PLANT DISEASE CONTROL

Crop	Disease	Suggested Dosage of Streptomycin Sulfate (ppm)	Remarks
Apple	Fireblight (*Erwinia amylovora*)	50–100	Apply when maximum temperatures above 18.3°C (65°F) occur during bloom, accompanied by or following rain. Where moderate to severe blight conditions prevail, use 100 ppm. Two applications are usually required 4–7 days apart between early bloom and petal fall
Celery	Bacterial Blight (*Pseudomonas apii*)	200	Use in plant bed only. Apply first spray when seedlings are in the 2-leaf stage. Reapply at 4–5 day intervals until transplanting time if blight is a problem
Chrysanthemum	Bacterial Blight (*Erwinia chrysanthemi*)	50	Soak cuttings in streptomycin solution for 4 hours immediately before rooting
Dieffenbachia	Bacterial Soft Rot (*Xanthomonas dieffenbachia*)	200	Soak cuttings in streptomycin solution for 20 min, and plant in sterilized soil
Pears	Fireblight (*Erwinia amylovora*)	100	Same as apple fireblight
Philodendron	Bacterial Leafspot	200	At first sign of watersoaked areas on leaves, apply spray and repeat at 4–5 day intervals as long as problem persists
Pepper Tomato	Bacterial Spot (*Xanthomonas vesicatorium*)	200	When seedlings are in 2-leaf stage, apply first spray of streptomycin and reapply at 4-day intervals until transplanting time if disease is a problem
Roses	Crown Gall (*Agrobacterium tumefaciens*)	200	Remove infected plants. Cut off galls if practical. Soak root system in streptomycin solution for 15 min. Replant in disease-free soil

		50	Following treatment with 200 ppm of streptomycin, apply 50 ppm streptomycin in watering solution or in foliar sprays at 7–10 day intervals for a 3-week period
Tobacco	Wildfire (*Pseudomonas tabaci*) Angular leafspot (*Pseudomonas angulata*)	200 or	Apply to plants in the plant bed starting when the plants are in the 2-leaf stage using a volume of 5 gal. per 300 ft of bed applied at 7-day intervals.
		100	Apply as a drench using a volume of 10 gal. per 300 ft of bed. Streptomycin may be combined with dithiocarbamate fungicides to provide control of fungus leaf-spot diseases

The agricultural uses of cycloheximide have been reviewed by Ford, Klomparens, and Hamner (1958).

The systemic nature of CHI (cycloheximide) has been demonstrated in several plant species. According to Starzk and Mitchell (1963), while no translocation of CHI was observed when it was applied to the primary leaves of corn, cowpea, and cucumber, the compound moved outward towards the growing tip of sunflower, and was distributed uniformly throughout the bean plant when applied to the roots and by foliar application to bean plants. Lemin and Magee (1957), in radioactive studies with cycloheximide acetate labelled with C^{14}, found that the compound was absorbed by tomato roots and induced antifungal properties in the leaves.

A number of formulations of cycloheximide are available commercially. These are summarized in Table 8.3.

TABLE 8.3. FORMULATIONS OF CYCLOHEXIMIDE

Product Trade Name	Active Ingredient	Suggested Uses
Acti-spray	380 mg tablets 7.7% cycloheximide	Cherry leafspot (*Coccomyces hiemalis*) Cherry powdery mildew (*Podosphaera oxyacanthae*) Cedar apple rust (*Gymnosporangium juniperi-virginianae*)
Acti-dione PM	0.027% cycloheximide	Rose powdery mildew (*Sphaerotheca pannosa* var: *rosae*) Rose rust (*Phragmidium* sp.) Powdery mildew of ornamentals (*Erysiphe cichoracearum*)
Acti-dione RZ	1.3% cycloheximide 75% PCNB	Azalea petal blight (*Ovulinia ozaleae*) Turfgrass diseases Large brown patch (*Rhizoctonia solani*) Melting-out (*Helminthosporium* spp.) Dollarspot (*Sclerotinia homeocarpa*) Rust (*Puccinia graminis*) Powdery mildew (*Erysiphe graminis*) Gray leafspot (*Piricularia grisea*) Grease spot (*Pythium* spp.) Helminthosporium leafspot (*Helminthosporium* spp.)

Cycloheximide has shown biocidal activity towards a number of soil-inhabiting fungi. The results of "in vitro" tests of the biocidal activity of CHI are summarized in Table 8.4.

TABLE 8.4. THE ACTIVITY OF CYCLOHEXIMIDE AGAINST SELECTED SOIL AND VASCULAR MICROORGANISMS

Microorganism	Dosage of Cycloheximide (ppm)	Bioactivity	Comments
Armillaria mellea	1–500	+++	20 ppm was lethal dose
Ceratostomella ulmi	200	+	
Endoconidiophora fagacearum	1	+++	Complete inhibition of germination of both ascospores and endoconidia
	8		Inhibition of growth. Fungistatic at 40 ppm. Completely fungicidal at 200 ppm
Fomes sp	5	+++	
Fusarium bulbigenum var. *niveum*	100	++	
F. lycopersici	100	+++	Complete inhibition at 100 ppm; Moderate activity at 50 ppm
	980	+++	
	100	++	
	100	+++	Complete inhibition of growth in 72 hours
F. solani	500	+++	Complete inhibition at 500 ppm. Partial inhibition at 100 ppm and lower
Fusarium spp.	5–100	+ to ++	Agar test. Activity varied with isolate.
	up to 3000	— to ++	Greenhouse test with bean. Activity varied with isolate. Phytoxic at 3000 ppm
Phytophthora cactorum	4	+++	Complete inhibition of growth
P. cinnamoni	5–25	+++	Fungicidal at 5 ppm as a dry soil mix, at 25 ppm in agar and as a soil drench
Pythium debaryanum	5; 10; 980	+++	
	4	+++	Complete inhibition of growth
	20	+++	Complete inhibition of growth in 72 hours
P. graminicola	1–100	+++	Complete inhibition at 100 ppm; severe retardation at 1 and 10 ppm
P. irregulare	1–5	+++	Highly inhibitory to growth. Little effect on sporangial germination in solution but highly inhibitory in soil
P. rostratum	1–5	+++	Inhibitory to growth and sporulation

TABLE 8.4 *(Continued)*

Microorganism	Dosage of Cycloheximide (ppm)	Bioactivity	Comments
P. ultimum	980 1–5	+++ + to +++	Growth inhibition varied with media nutrients and isolate. Strongly inhibitory to sporulation
Pythium spp.	10–100	+++	
Rhizobium spp.	980 20 and 100	—	No activity against *R. meliloti* and *R. trifolii*. *R. trifolii* no nodulation of subterranean clover at 100 ppm; reduced and delayed nodulation at 20 ppm. No effects in isolation media at 50 ppm
Rhizoctonia solani	50 10	+++ +++	Complete inhibition at 50 ppm; severe retardation at 1 and 10 ppm.
	100	+++	Stem infection of potato reduced by watering with 100 ppm
	100	+++	Complete inhibition of growth
	1.25	+++	Complete inhibition of growth in 72 hours
Sclerotium rolfsii	25	+++	Complete prevention of mycelial growth
	2.5	+++	Complete inhibition of growth in 72 hours
Thielaviopsis basicola	4	+++	Complete inhibition of growth
Trichoderma viride	100	+++	Complete inhibition at 100 ppm. Partial inhibition at 20 ppm
Verticillium albo-atrum	500	+++	Fungicidal in soil drench test

Most studies were in vitro unless indicated otherwise.
+++ = high biocidal activity.

The toxicology of cycloheximide has been extensively investigated and has been summarized in a comprehensive review by the Upjohn Company (Anon. 1975). The following summary of the mammalian toxicity and environmental hazards of Acti-dione products is taken directly from the Upjohn Company literature.

Toxicity, potential health hazards, and the impact of pesticides on the environment are subjects that have been widely discussed and reported in news media during the past few years. One of the most common measurements of pesticide toxicity is the Acute Oral LD_{50} to the rat. This test involves oral administration of the test chemical at various levels to laboratory rats to establish the dose required to kill 50% of the test animals. The LD_{50} is expressed in mg/kg, meaning the milligrams of test chemical per kilogram of body weight required to kill 50% of the animals. Most reported LD_{50} values refer to technical or pure chemical as opposed to formulated commercial products.

Two important points must be considered when evaluating the LD_{50} value for Acti-dione. First, it must be clearly noted whether pure technical chemical or a more dilute formulated product was used in the test. Second, we must consider the wide species variation in the response of test animals to Acti-dione. The following comments relate to these two subjects.

Cycloheximide, the active ingredient in Acti-dione fungicides, has an Acute Oral LD_{50} of 1.75 mg/kg in rats. Unfortunately such data are misleading and do not present the true picture regarding the toxicity and possible health hazards of formulated Acti-dione products. When evaluating potential hazards to the consumer, primary consideration should be given the formulated products actually being used—not the technical cycloheximide. Because of its very potent fungicidal properties the concentration of cycloheximide in formulated Acti-dione products ranges from only 0.027% in Acti-dione PM, to a maximum of 2.1% in Acti-dione TGF. The Acute Oral LD_{50} values of these various Acti-dione products in rats are compared to technical cycloheximide as follows:

Product	% Cycloheximide	Acute Oral LD_{50} (mg/kg) Rat
Technical Acti-dione	95	1.75
Acti-dione PM	0.027	5,500
Acti-dione Thiram	0.75	189
Acti-dione RZ	1.3	217
Acti-dione TGF	2.1	156

It should be noted that Acti-dione Thiram and Acti-dione RZ are combination products with tetramethylthiuram disulfide and pentachloronitrobenzene, respectively, and therefore reflect the toxicity of these fungicides as well as that of cycloheximide.

A second consideration in attempting to estimate possible hazards of Acti-dione to the user based on Acute Oral LD_{50} data in rats, is the dramatic species variation in response to this chemical. While the Acute Oral LD_{50} in rats is 1.75 mg/kg, the analagous value in mice is 133 mg/kg. This means that the rat is about 75 times more sensitive to cycloheximide than the mouse. Since cycloheximide was originally evaluated as a potential human drug, evidence is plentiful to indicate that man is one of the more tolerant species. Single doses of greater than 700 mg. have been administered successfully to humans. Human beings with fungal diseases have tolerated 30 mg. four times daily intramuscularly and 180 mg. per day intravenously. Cycloheximide has been used in clinical studies for cancer therapy in doses up to 15 mg/kg two to three times per week up to 48 days.

Acti-dione (cycloheximide) was discovered in 1946; and since that time, has been widely used for plant disease control. In more than 20 years of commercial use, there has never been a single case of authenticated accidental poisoning. Although this is testimony to the relative safety of Acti-dione products, it is important to remember that as with any pesticide, they should be used in accordance with label directions, and handling and use precautions should be carefully noted.

OTHER ANTIBIOTICS FOR PLANT DISEASE CONTROL

Although streptomycin and cycloheximide are the most widely used antibiotics for plant disease control, a number of other oxytetracycline and chlorotetracycline derivatives have been used at concentrations of 25–50 ppm for the control of postharvest bacterial soft rot of spinach and lettuce. They have been reported to be more effective for this purpose than streptomycin. They have also given control of potato soft rots in laboratory trials.

The antifungal polyene antibiotics, nystatin and pimaricin, have shown promise for the postharvest control of peach rots, strawberry rots, Gloeosporium rot of banana, and Botrytis infection of roses and orchids in storage. Nystatin and pimaricin have been reported to be strongly biocidal to most phytopathogenic fungi except those belonging to the Class Oomycetes.

Other antibiotics that have been considered for the control of several plant diseases are Griseofulvin, Clavacin, Chloromycetin, Neomycin, Aureomycin (chlorotetracycline), Phytoactin, Terramycin (oxytetracycline), and Phytomycin (streptomycin nitrate).

BIBLIOGRAPHY

ANON. 1975. Toxicity of Cycloheximide. A review. The Upjohn Company, Kalamazoo, Mich.

ARK, P.A. 1949. Streptomycin. Williams and Wilkins Co., Baltimore.

ARK, P.A., and S.M. ALCORN. 1956. Antibiotics as bactericides and fungicides against diseases of plants. Plant Dis. Rep. 40: 85.

CHAIN, E. *et al.* 1940. Penicillin as a chematherapeutic agent. Lancet 11: 226–228.

DAVIS, S.H., R.E. ENGEL, and G. SILBER. 1951. Control of Brown Patch of turf in New Jersey. Phytopathology 41: 657.

FELBER, I.M., and C.L. HAMNER. 1948. Control of powdery mildew on bean plants by means of an antibiotic. Bot. Gaz. 110: 324.

FLEMING, A. 1929. On the antibacterial action of cultures of Penicillin, with special reference to their use in the isolation of *B. influenzae*. Br. J. Expt. Pathol. 10: 226–236.

FLOREY, H.W. 1949. Historical introduction to antibiotics, Volume 1. Oxford University Press, London.

FORD, J.H., W. KLOMPARENS, and C.L. HAMNER. 1958. Cycloheximide (Acti-Dione) and its agricultural uses. Plant Dis. Rep. 42 (5)680–695.

GOTTLIEB, D., and P.D. SHAW. 1970. Antibiotics. Ann. Rev. Phytopathol. 8: 371–402.

LEACH, B.E., J.H. FORD, and A.J. WHIFFEN. 1948. Acti-Dione. An antibiotic from *Streptomyces griseus*. J. Am. Chem. Soc. 70: 1223–1225.

LEBEN, C., and G.W. KEITT. 1956. Antibiotics and plant disease. Agric. Food Chem. 2 (5)234–239.

LEMIN, A.J., and W.E. MAGEE. 1957. Degradation of cycloheximide derivatives in plants. Plant Dis. Rep. 41: 447.

MOSS, V.D. 1961. Antibiotics for control of Blister Rust of western white pine. For. Sci. 7 (4)380–396.

MURNEEK, A.E. 1952. Cycloheximide for the control of fireblight of apples. Phytopathology 42: 57.

PETERSON, D., and D. CATION. 1950. Exploratory experiments in the use of Acti-Dione for the control of Peach Brown Rot and Cherry Leafspot. Plant Dis. Rep. 34: 5.

POTTER, M.C. 1908. On a method of checking parasitic diseases in plants. J. Agric. Sci. 3: 102–107.

SANFORD, G.B. 1926. Some factors affecting the pathogenicity of *Actinomyces scabies*. Phytopathology 16: 525–547.

SCHATZ, A., E. BUGIE, and S.A. WAKSMAN. 1944. Streptomycin, a substance exhibiting antibiotic activity against gram positive and gram negative bacteria. Proc. Soc. Exp. Biol. Med. 55: 66–69.

STARZK, M.J., and J.E. MITCHELL. 1963. Cycloheximide behaviour in bean and cherry tissue. Phytopathology 53: 309–312.

TORGESON, D.C. 1967. Fungicides: an Advanced Treatise, Volume 1. Academic Press, New York.

VAUGHAN, J.R. 1951. Cycloheximide, an antibiotic effective against turf diseases. Phytopathology 41: 36.

WHIFFEN, A.J. 1948. The production, assay, and antibiotic activity of Acti-Dione—an antibiotic from *Streptomyces griseus*. J. Bacteriol. 56: 283–291.

WHIFFEN, A.J. 1946. The production of an antifungal antibiotic from *Streptomyces griseus*. J. Bacteriol. 52: 610–611.

WHIFFEN, A.J. 1950. The activity 'in vitro' of cycloheximide (Acti-Dione) against fungi pathogenic to plants. Mycologia 42: 253.

9

Laboratory Bioassay of Chemicals for Plant Disease Control

Within the past twenty-five years a host of ghostly chemicals has paraded across the stage of agricultural progress as potential candidates for the honor of "All American Blight Buster." Most of these never got off the ground, a few flamed brightly for a brief period and then were extinguished, some served a short term as useful plant protectants only to be replaced by more effective compounds, and still others were purged by new pesticide regulations. Only a very few chemical compounds have survived the competition to become widely accepted as chemicals for plant protection. In the years ahead still other compounds will join the parade; some will fail, a few will be promising, and very few will make the grade as accepted partners in the recognized practices for plant disease control. The selection, development, and introduction of candidate fungicides have been handicapped by lack of accepted laboratory testing methods. Separating successful compounds from ineffective materials for plant disease control is an expensive and time-consuming procedure. It has been estimated that the development of one new successful fungicide costs the agricultural chemical industry $3 million before the fungicide is offered for commercial distribution.

Prior to the development of modern methods for fungicide bioassay two major difficulties arose. One of the most persistent of these problems was the difficulty encountered in duplicating the results of preliminary laboratory tests. The other puzzling problem was the difficulty of predicting field performance on the basis of laboratory experiments. Efforts of the American Phytopathological Society, working in cooperation with major agricultural chemical companies, have been primarily responsible for the acceptance of modern methods for fungicide bioassay. Modern methods for screening fungicides in the laboratory now permit reasonably reliable reproduction of results from day to day, and between research workers conducting independent tests. Advanced quantitative procedures and modern methodology now make it possible to predict with some accuracy the protective value of candidate fungicides under field conditions.

The purpose of the following discussion is not to equate the merits or disadvantages of "in vitro" bioassay of fungicides, but to provide guidelines for those interested in the methodology of fungicide bioassay. The literature on "in vitro" testing of fungicides is extensive, and has been adequately reviewed by Horsfall (1945), McCallan (1947), and Torgeson (1967).

Accepted methods for fungicide bioassay yield results rapidly, and eliminate the need for costly testing of compounds in the field by screening out ineffective chemicals in the laboratory. Laboratory bioassay of fungicides does not take into consideration the host-parasite-chemical relationship and is thus limited.

"In vitro" studies of candidate fungicides do provide reliable information on fungitoxicity against specific phytopathogens and may be used for preliminary consideration of the compatibility of the candidate fungicide with standard insecticides. The results of "in vitro" laboratory studies of candidate compounds serve as a reliable basis for further "in vivo" and field testing procedures.

The reliability of "in vitro" tests for fungitoxicity have been questioned by some plant pathologists. Marsh, in 1936, and Howard, in 1939, compared spore germination on leaves and on glass slides and obtained comparable results. Miller, in 1943, compared the retention of copper on leaves and on pyralin plates and found laboratory methods to be reasonably accurate. "In vitro" fungicide bioassay should be used only for preliminary measurement of fungitoxicity, and may also be utilized for preliminary determination of tenacity, stability, and redistribution properties. Compounds toxic to fungi may affect them in various ways. They may inhibit growth of mycelium, spores may plasmolyze and fail to germinate, or spores may be permanently inactivated. A fungicide has been defined as "any substance that destroys or inhibits the growth of fungous spores and mycelium." Most plant pathologists restrict the term fungicide to those compounds that actually kill the test fungus. The term "fungistat" is usually applied to those compounds that inhibit growth or germination without actually killing the test fungus.

There are five methods of procedure for the laboratory testing of candidate fungicides "in vitro," and these may be categorized as follows:

a. The depression-slide method for fungicide bioassay
b. The slide-germination technique for fungicide bioassay
c. The paper disk method for fungicide bioassay

d. Bioassay of eradicant fungicides in the laboratory
e. Miscellaneous methods

All of the above methods require the preparation of standardized spore suspensions of the test fungus (or fungi), and the preparation of stock solutions of the test compounds.

STANDARDIZED PROCEDURE FOR THE PREPARATION OF CALIBRATED SPORE SUSPENSIONS

Standardized spore suspensions of the test fungi should be prepared *not more than 30 min* before setting up the bioassay test. The test fungi should be grown in petri dish culture using suitable media such as potato dextrose agar or malt agar, and cultures of specified age, usually ten days old. The following procedure is suggested:

a. Flood the petri dish containing the test fungus with sterilized distilled water.
b. Scrape the surface of the culture with a sterilized "rubber policeman," raising the cover of the petri dish sufficiently to insert the scraper.
c. Decant the spore-mycelium suspension through a double layer of sterilized cheesecloth into a sterilized 150 ml erlenmeyer flask.
d. Determine the approximate number of spores per milliliter of the suspension, using a blood-counting cell under the microscope.
e. Place a drop of the spore suspension on each side of a hemacytometer (blood-counting cell) at the depression points.
f. Place the blood-counting cell under the low power of a microscope (10 X × 10 X) and count the number of spores in five squares on each half of the slide. The five squares should be selected in an "X" pattern.
g. The total number of spores counted in the designated squares multiplied by 1000 equals the number of spores per milliliter of spore suspension (e.g., a total spore count of 100 is equivalent to 100,000 spores per ml). (Refer to Tuite (1969) for details on determination of spore concentration.)
h. Adjust the concentration of spores in the suspension to 50,000 spores per milliliter by dilution with distilled water.

STANDARD PROCEDURE FOR PREPARATION OF TEST-CHEMICAL STOCK SOLUTION

The concentration of the chemical test solution is expressed in *parts per million* (ppm) and is calculated by the following formula:

$$\frac{\text{X units of solute}}{\text{1,000,000 units of solvent}} = \frac{\text{Y grams}}{\text{1,000,000 ml of water}}$$

thus for "X" ppm in "Y" ml of solvent

$$\frac{\text{X grams of chemical}}{\text{1,000,000 ml of water}} = \frac{\text{Z grams}}{\text{Y ml of water}}$$

For example: It is desired to calculate the number of grams of chemical needed to make a 500 ppm solution in 100 ml of water. Thus:

$$\frac{\text{500 grams of chemical}}{\text{1,000,000 ml of water}} = \frac{\text{"Z" grams}}{\text{100 ml water}}$$

Solving for "Z" = 0.05 grams

This means that 0.05 grams of chemical in 100 ml water will give a 500 ppm solution, providing the chemical used is 100% active ingredient. However, most commercial formulations of fungicidal chemicals are not 100% active ingredient. This means that a correction factor must be used to arrive at the amount of chemical required on the basis of 100% active ingredient.

The correction factor is calculated as follows:

If the test chemical contains 50% active ingredient, the correction factor becomes:

$$\frac{100}{50} = 2X$$

This means that the grams of chemical needed to give a specific concentration calculated by the first formula must be multiplied by "2."

Some common correction factors for fungicides containing different active ingredient content are as follows:

Active ingredient concentration	Multiply by
80%	1.2500
76%	1.3158
75%	1.3333

70%	1.4286
66%	1.5152
65%	1.5385
60%	1.6667
55%	1.8182
50%	2.0000
25%	4.0000
22%	4.5454
20%	5.0000

THE SLIDE-GERMINATION FUNGICIDE BIOASSAY TECHNIQUE

Slide-germination methods of bioassay using depression slides and standard microscope slides for screening compounds for possible fungicidal value provide a rapid method of evaluation. The two techniques, depression slide and standard slides, yield results within 24 hours, and the experiment may be set up in the laboratory in a relatively short time. One of the limitations of these techniques is their unsuitability for fungi that do not sporulate readily, or are difficult to culture on artificial media. Such fungi are not suited to the bioassay technique discussed below.

The glass depression-slide technique has one advantage over the sprayed slide technique in that it does not require calibrated equipment for the application of the test chemical. The chemical being tested and the spores of the test fungus are incorporated in the same solution. This means that chemical and spores are in continual contact with each other, and consequently the test does not simulate conditions as they might exist under field conditions. Furthermore, a fungus spore that requires oxygen for its maximum germination may be inhibited by lack of oxygen in the bottom of the well of a depression slide. Fungicide tests on depression-slides are also hampered by the tendency of spores to aggregate, making the recording of germination difficult.

In the sprayed microscopic slide method of bioassay the test chemical is sprayed on the surface of the slide using precision application equipment, and the sprayed slide is then allowed to dry. Once the sprayed slide has dried, a drop of spore suspension of the test fungus is placed on the sprayed surface. The building and calibration of the necessary precision spray equipment is expensive and difficult, and the technique is more time-consuming than the depression slide method.

STANDARD PROCEDURE FOR DEPRESSION SLIDE BIOASSAY

For the bioassay of five chemicals at three concentrations (125, 250 and 500 ppm) and one check slide against one test fungus, the following equipment is necessary:

16 – 10 ml test tubes
1 – test tube rack
16 – depression slides (3 in. × 1 in.)
16 – 100 ml beakers
16 – 1000 ml beakers
16 – large (5½ in. diameter) petri dishes
16 – filter paper disks
16 – strips of blotting paper
10 – 10 ml pipettes
10 – 50 ml erlenmeyer flasks
6 – 250 ml erlenmeyer flasks containing 100 ml distilled water
2 – 2 ml pipettes
1 – 125 ml erlenmeyer flask

Cleaning Glassware

Thorough cleaning of all glassware used in any of the bioassay tests is extremely important. All glassware should be thoroughly cleansed with sodium dichromate-sulfuric acid cleaning solution (see Tuite (1969)). After cleaning, rinse five times in tap water followed by one rinse in clean distilled water. Following cleaning and rinsing all glassware should be sterilized in an autoclave at 15 lb pressure for 30 min. The depression slides should then be coated with cellulose-nitrate to provide a uniform surface in the depression well for all depression slides. Cellulose-nitrate may be purchased from most scientific chemical supply firms under the trade name "Pyroxylin." It should be dissolved in butyl acetate to give a 2.5% solution. The depression slides should be dipped in this preparation and dried overnight in a dust-free room.

Preparation of Stock Solution

A stock solution of the test chemicals should now be prepared as follows:

Make up 100 ml of each of the test chemicals at twice the desired concentration; viz. since the highest desired concentration in this example is 500 ppm, make up a stock solution for each chemical at a concentration of *1,000* ppm. MARK THE STOCK SOLUTION 500 PPM, since it will be diluted later to this concentration when the spore suspension solution of the test fungus is added to it.

Preparation of Moist Chambers

Special moist chambers will be required to store the depression slides containing test chemical and fungus spore suspension. These can be prepared by filling 16 large petri dishes half full of water, and then placing a 100-ml erlenmeyer flask upside-down in the middle of the water reservoirs. Line the bottom of 1,000-ml beakers with a filter paper disk and line the sides with a strip of blotting paper. The prepared depression slides will later be placed on top of the small beaker and the whole covered with the large, lined beaker (Fig. 9.1).

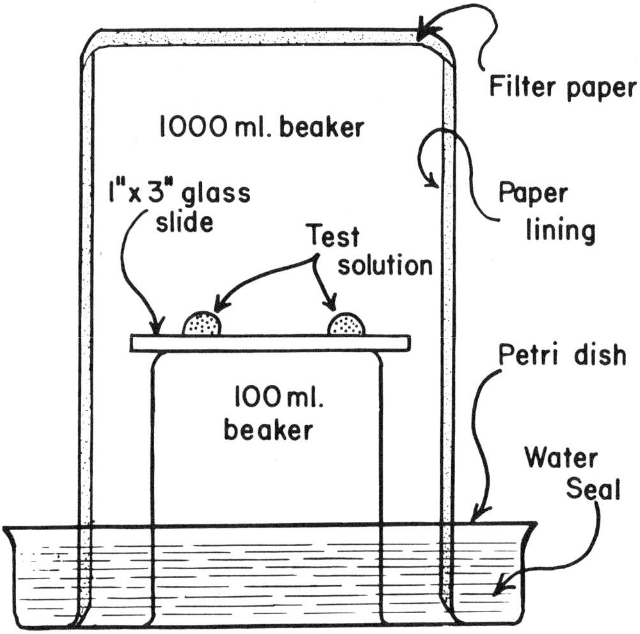

FIG. 9.1. MOIST-CHAMBER INCUBATOR FOR SLIDE GERMINATION BIOASSAY FUNGICIDE TESTS

Starting the Bioassay Test

(Aseptic precautions must be used in all of the steps of the depression slide bioassay method.)

a. Swab laboratory bench on which the experiment is to be set up with hyamine (or other general disinfectant) and place cheesecloth saturated with hyamine on swabbed laboratory bench.

b. Prepare a dilution series for each test chemical as follows:
   1. Place 10 ml of sterilized distilled water in each of two 50-ml flasks and mark one of them 250 ppm, and the other 125 ppm. (Do this for each test chemical.)
   2. Using a sterilized 10 ml pipette, place 10 ml of the prepared stock solution into the flask marked "250 ppm." Mix thoroughly with the water already in the flask. This is the 250 ppm concentration of the test chemical.
   3. Using a pipette, remove 10 ml of the solution from the flask marked "250 ppm" and place it in the flask marked "125 ppm," which already contains 10 ml of sterilized distilled water. This is now the 125 ppm concentration of the test chemical.
   4. The dilution series is now complete, viz. the stock solution which is 500 ppm, the flask from (2) which is 250 ppm, and the flask from (3) which is 125 ppm.
   5. Using a sterilized 1-ml pipette, now transfer 1 ml of each concentration to a test tube marked with the name of the test chemical and the concentration of the solution. Place these in a test tube rack. You will now have three test tubes marked 500 ppm, 250 ppm, and 125 ppm for each test chemical.
   6. Place 1 ml of sterilized distilled water in another test tube and mark the tube "Check."
   7. Set up a dilution for each of the test chemicals exactly as outlined above. Other concentrations may be prepared using the same dilution procedure.

Preparation of the Spore Suspension

Prepare a calibrated spore suspension of the test fungus as previously described. Using 10 ml pipettes place 1 drop of standardized spore suspension in each well of the depression slides. To each of the depression slide wells now add one drop from the test tubes containing the test chemical. Mark slide with the chemical designation and concentration. The depression slides will now contain one drop of standardized spore suspension and one drop of test chemical at the desired concentrations of 500, 250, and 125 ppm. Immediately place the prepared depression slides in the moist-chamber incubators. Record spore germination in each of the preparations under the low power of the microscope at 24-hour intervals until there is no further increase in percentage germination. The completed procedure for the depression slide method is illustrated in Fig. 9.2.

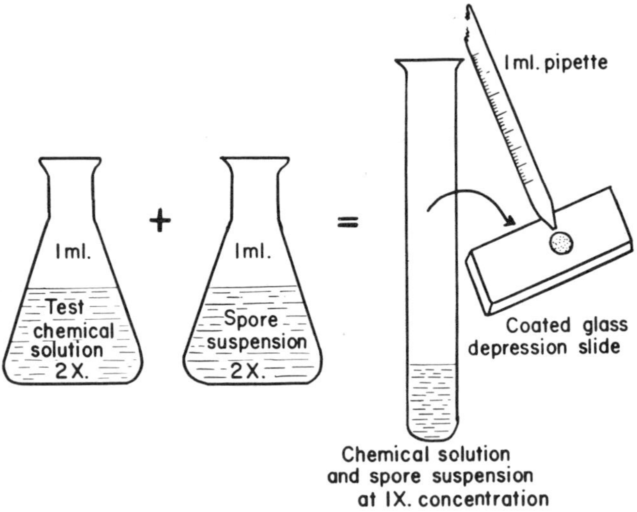

FIG. 9.2. LABORATORY PROCEDURE FOR THE DEPRESSION SLIDE METHOD OF FUNGICIDE BIOASSAY

STANDARD PROCEDURE FOR THE SLIDE-GERMINATION METHOD OF FUNGICIDE BIOASSAY

For the bioassay of five chemicals, at three concentrations, and one Check, against one test fungus, the following equipment is necessary:

16 – glass slides (3 in. × 1 in.)
16 – 100 ml beakers
16 – 1000 ml beakers
16 – large petri dishes (5½ in. diameter)
16 – filter paper disks
16 – strips of blotting paper
5 – 500 ml erlenmeyer flasks containing 250 ml distilled water
10 – 250 ml erlenmeyer flasks containing 100 ml distilled water
1 – 125 ml erlenmeyer flask containing 50 ml distilled water
2 – 2 ml pipettes
1 – horizontal precision laboratory sprayer

Cleaning Glassware

All glassware should be thoroughly cleaned with sodium dichromate-sulfuric acid cleaning solution. After cleaning, rinse in tap water five times and then in clean distilled water. All glassware should then be sterilized in an autoclave at 15 lb pressure for 30 min.

Coating Glass Slides

Since it is almost impossible to obtain a uniform glass surface by cleaning and rinsing, all of the microscope slides should be coated with cellulose nitrate, which will provide a consistently uniform surface. Cellulose nitrate is available under the trade name "Pyroxylin" from several chemical supply companies. The pyroxylin should be dissolved in butyl acetate to provide a 2.5% solution. The slides should be dipped in this solution, and hung up to dry on a suitable rack in a dust-free room. All glass slides should bear a serial number etched on the glass surface with a diamond point or other tool.

Preparation of Stock Solution

Stock solutions of each of the test chemicals should be prepared containing 500 ppm of the test chemical in 250 ml of distilled water in 500-ml erlenmeyer flasks. The stock solution should be adjusted for active ingredient content and the stock solution should contain 500 ppm on the basis of 100% active ingredient. The desired concentations of 250 ppm and 125 ppm are prepared by simple dilution of the stock solution of the test chemical. With some highly fungicidal compounds much lower concentrations of test chemical may be desirable and may be prepared by standard dilution methods.

Application of Fungicide

The various concentrations of the test chemicals are applied to coated, marked slides by means of a calibrated, precision laboratory sprayer (Fig. 9.3). For a detailed description of this equipment, refer to Horsfall, Heuberger, Sharvelle, and Hamilton (1940). Using this equipment a uniform, replicable dosage of the test chemicals is applied to the glass slides which should be allowed to dry after the spray application.

Preparation of Sprayed Slides

After the coated glass slides have received a uniform deposit of each concentration of each of the test chemicals and have dried, they are placed in the moist-chamber incubators as previously described. Now, using a 2-ml pipette two separate drops of the calibrated spore sus-

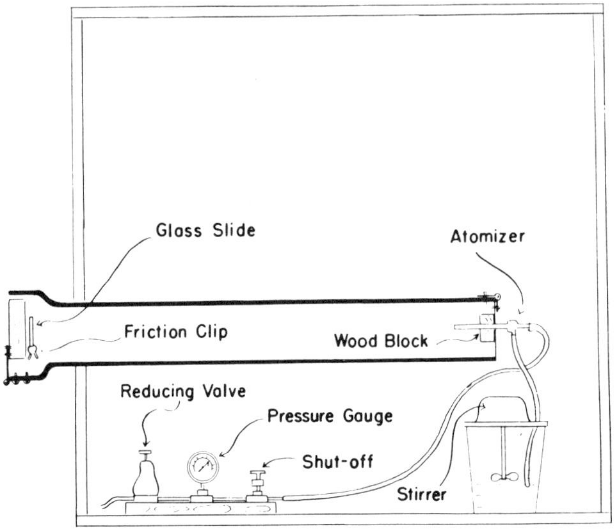

FIG. 9.3. THE HORIZONTAL LABORATORY PRECISION SPRAYER

pension are placed on each of the sprayed slides. An unsprayed slide or a slide sprayed with distilled water and allowed to dry is set up as a "check slide" or "control" for spore germination tests. The percentage spore germination of the test fungus on each of the sprayed slides is then determined and recorded at 24-hour intervals. Germination counts usually may be discontinued after 48 hours.

The use of a double counting tally will expedite the laborious task of counting spore germination. An arbitrary criterion for deciding whether or not a spore has germinated is that if the sides of the germ tube are parallel, or if the germ tube exceeds in length more than half the diameter of the spore, it is considered to have germinated.

Interpretation of Results

In most slide-germination bioassay trials three replicated slides for each concentration of the test chemicals are desirable. Sufficient spore germination data should be obtained to permit plotting a dosage-response curve on logarithmic graph paper. For each curve the LD_{50} (lethal dose 50% is the chemical dosage that will inhibit 50% of the germination) may be determined.

The "Bordeaux coefficient" has been used by several investigators as a standard of comparison of fungicide toxicity. Any other standard fungicide could also be used.

Phenol has been used as a standard of effectiveness for bactericides. Copper sulfate and bordeaux mixture have been used as comparative standards for fungicides. Ratios of LD_{50} values of standard and test material give a rough approximation of the relative performance of the candidate compound.

The Bordeaux coefficient (comparative coefficient) may be determined by the following formula:

$$\text{Bordeaux coefficient or Comparative coefficient} = \frac{LD_{50}\text{ value for bordeaux mixture (or other fungicide) in parts per million}}{LD_{50}\text{ value for candidate compound}}$$

Using this formula, the fungitoxicity of the candidate compound may be directly compared with the fungitoxicity of Benlate, Captan, Maneb, or other fungicide.

A number of problems may be encountered with the slide-germination bioassay technique. One of the most frequent problems is the difficulty of maintaining uniform droplet size on the glass slides. Several suggestions have been offered to correct this difficulty. It has been suggested that inscribing circles on the slides with a diamond point, or drawing uniform circles with a wax pencil would eliminate this difficulty. Others have suggested etching circles with hydrofluoric acid. In the many fungitoxicity trials conducted by the author in the Horsfall laboratory, the use of a cellulose acetate coating was the simplest and most effective method for ensuring uniform droplet sizes. It has also been observed that most fungus spores germinate more uniformly on cellulose nitrate-coated slides. However, some fungus spores do not germinate well in liquid droplets. Age of spores influences viability and in general, the older the spores the more resistant they are to fungitoxic action. A standard spore age of 10 days has been adopted for test fungi used in slide-germination fungicide tests. Fungus spores that do not consistently germinate 50% or more should not be used in fungicide toxicity tests. Vigorous shaking of spore suspensions of the test fungus before application to the glass slides will generally improve the viability of spores of the test fungi.

STANDARD PROCEDURE FOR THE MODIFIED PAPER DISK METHOD OF FUNGICIDE BIOASSAY

The modified paper disk method (Sharvelle and Pelletier 1956) is especially designed for use with fungi that do not sporulate readily in artificial culture. This method is adapted from bioassay procedure used in the assay of bactericides, and is slower than other procedures requiring 10–14 days for the completion of the test but does not require the use of a microscope and the tedious procedure of counting spore germination.

Preparation of Agar

a. Peel and slice 200 grams of potatoes in 500 ml water.
b. Add 10 grams of agar to above.
c. Cook agar and potatoes for 30 min in laboratory steamer.
d. After cooking strain through cheesecloth.
e. Add 10 grams of dextrose to the strained agar-potato mixture.
f. Sterilize in autoclave for 30 min at 15 lb pressure.

Preparation of Paper Disk Plates

Place a square piece of blotting paper in a standard petri dish so that the four corners touch the sides of the dish. Place six 12.7 mm Schleicher and Schnell penicillin bioassay disks on the blotting paper in two rows of three each. (Paper disks are available from most scientific supply houses.) Prepare as many paper disk plates as are necessary for the test.

Pouring the Agar Plates

Agar plates should now be poured using pre-cooled agar at 40°C in sterilized standard petri dishes. A 250-ml erlenmeyer flask containing 150 ml of agar will pour approximately 12 plates. After the plates cool mark all of them with suitable indication of the test chemical and concentration to avoid later confusion.

Preparation of Test Chemical Stock Solutions

Prepare stock solutions of the candidate fungicide at the desired concentrations using the dilution procedure previously described.

Glassware Needed for the Bioassay

The following equipment will be required for the bioassay of five test chemicals at three concentrations against one test fungus:

17 – paper disk plates
34 – poured agar plates
5 – 250-ml erlenmeyer flasks
1 – 125-ml erlenmeyer flask

11 – 50-ml erlenmeyer flasks
1 – 2-ml pipette
10 – 10-ml pipettes
1 – glass stirring rod with rubber policeman.

All of the above glassware should be sterilized in the autoclave.

Setting up the Paper Disk Bioassay

a. Starting with the 125 ppm dilution of the test chemical (or lowest dilution), impregnate the 6 paper disks previously prepared by adding 2 drops of the test chemical dilution to each disk using a 10-ml dropping pipette. Now proceed to the 250 ppm dilution and then to the 500 ppm dilution using the same procedure and the same pipette.
b. Proceed to the next test chemical and repeat the above procedure.
c. Prepare a suspension of the test fungus in sterilized distilled water by either using a calibrated spore suspension or a mycelial suspension prepared with a Waring blender using a petri dish culture of the test fungus if it is a "non-sporulator."
d. After all of the paper disks have been impregnated with the test chemical, seed each one with 1 drop of spore suspension or mycelial suspension of the test fungus using a 2-ml dropping pipette. The drop should be placed in the center of each impregnated disk without splashing.
e. Now transfer three impregnated-seed paper disks aseptically to each of the two poured agar plates. A series of check disks (impregnated with sterilized distilled water instead of test chemical) should be set up in a similar manner.
f. Store impregnated-seed paper disks on agar in petri dishes at the desired temperature in the laboratory, and observe for growth of the test fungus.

Recording Results

The results for this method of fungicide bioassay are recorded as either positive growth (+), or negative growth (–) for each concentration of the chemicals in the bioassay. Observations should be recorded at 48-hour intervals over a 7–10 day period. From these observations the m.i.c. (minimum inhibition concentration) may be recorded. The m.i.c. is the chemical dosage at which no growth occurs from the impregnated disks over the 7–10 day test period.

The procedure for the modified paper disk fungicide bioassay method is illustrated in Fig. 9.4.

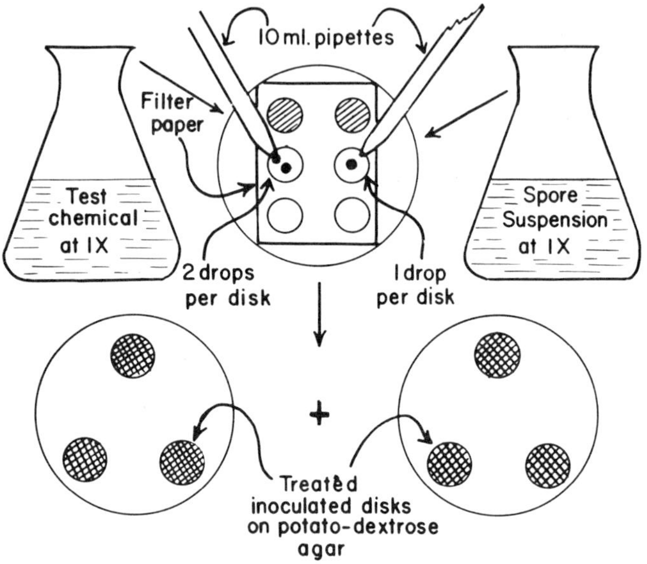

FIG. 9.4. THE MODIFIED PAPER DISK METHOD FOR FUNGICIDE BIOASSAY

PROCEDURE FOR BIOASSAY OF ERADICANT ACTION OF FUNGICIDAL CHEMICALS

It is frequently desirable to obtain some information in the laboratory relative to the ability of a candidate fungicide to eradicate or inhibit growth of test fungi. The following procedure may be used to make a preliminary bioassay of potential eradicant fungicidal properties.

Preparation of Inoculum

The test fungus is grown in petri dishes on malt agar for 10–14 days at a temperature favorable for growth.

Preparation of Bioassay Plates

a. The test chemicals are prepared as aqueous suspensions at 2X concentration (double the dosage required) using the procedure previously described for the preparation of test chemical stock solutions.
b. Double-strength malt agar is prepared using standard laboratory procedure for media preparation. Double all materials specified.
c. Mix equal volumes of (a) and (b) and pour into standard petri dishes using aseptic precautions.

Inoculation of Petri Dishes

a. Using an appropriate tool, cut 5-mm disks from the test fungus growing in petri dish culture, making sure the disks are of uniform thickness.
b. Place six disks with the fungus-bearing surface uppermost on the surface of the chemical-agar plates.

Recording of Results

After the fungus disks have been incubated for 14 days at the desired controlled temperature, transfer any disks showing no growth to the surface of fresh malt agar in petri dishes. Incubate for 21 additional days and record presence or absence of growth.

OTHER MACROSCOPIC TESTS FOR FUNGITOXICITY

Various other macroscopic tests of fungitoxicity have been used by several investigators. As early as 1927 Lee and Martin, and later in 1940 Wellman and Heald suggested placing known volumes of fungus spores in adjusted aqueous concentrations of toxicants for varying periods of time before transferring known aliquots to agar medium. The criterion of fungitoxicity with this method was reduction or inhibition of the fungus colonies that developed.

Mason and Powell (1947) suggested growing mycelium in nutrient culture containing known concentrations of the toxicants to be tested, and then weighing the resultant mycelial growth after a specific period of incubation under controlled conditions.

A bioassay method for fungi that fail to sporulate, or have limited spore production, was suggested by Carpenter (1942). This method consisted of incorporating the toxicants with agar and then seeding poured plates with uniform mycelial colonies in agar blocks. The zone of fungus inhibition was then used as the criterion for fungitoxicity.

Macroscopic bioassay tests for fungitoxicity generally are not as reproducible as are spore germination methods.

PREDICTING FUNGICIDE FIELD PERFORMANCE ON THE BASIS OF LABORATORY BIOASSAY

Horsfall and his co-workers (Horsfall, Heuberger, Sharvelle, and Hamilton 1940) were among the first to suggest that the protective value of fungicides in the field could be predicted on the basis of laboratory bioassay tests. Horsfall considered that fungitoxicity and tenacity were the two fundamental components of potential protective value against plant diseases in the field.

Waggoner *et al.* (Waggoner, Rich, and Horsfall 1952) have stated that a statistical prediction of fungicidal field performance could be made on the basis of toxicity, tenacity, and stability of the candidate fungicide in the laboratory. Most plant pathologists would hesitate to predict fungicide field performance on the basis of laboratory "in vitro" bioassay tests. Nevertheless, "in vitro" bioassay tests often correlate well with field performance and can serve as accelerated, relatively inexpensive tests for the selection of candidate fungicides worthy of further greenhouse and field tests.

BIBLIOGRAPHY

CARPENTER, J.B. 1942. A toximetric study of some eradicant fungicides. Phytopathology 32: 845–856.

COMMITTEE ON STANDARDIZATION OF FUNGICIDE TESTS (The American Phytopathological Society). 1943. The slide germination method of evaluating protectant fungicides. Phytopathology 33: 627–632.

DIMOND, A.E., J.W. HEUBERGER, and E.M. STODDARD. 1941. The role of the dosage response curve in the evaluation of fungicides. Conn. Agric. Exp. Stn. Bull. *451*.

EVANS, A.C., and H. MARTIN. 1935. The laboratory evaluation of water-soluble wetting agents as constituents of combined washes. J. Pomol. Hort. Sci. 13: 261–292.

HEUBERGER, J.W. 1940. A laboratory bioassay of tenacity of fungicides. Phytopathology 30: 545–563.

HORSFALL, J.G. 1945. Quantitative bioassay of fungicides in the laboratory. Bot. Rev. 11: 357–397.

HORSFALL, J.G., J.W. HEUBERGER, E.G. SHARVELLE, and J.M. HAMILTON. 1940. A design for laboratory assay of fungicides. Phytopathology 30: 545–563.

HOWARD, F.L. 1939. The value of testing fungicides in the laboratory before use in the field. Proc. Am. Soc. Hort. Sci. 37: 409–414.

LEE, A., and J.P. MARTIN. 1927. A method for testing "in vitro" the toxicity of dust fungicides to fungous spores. Phytopathology 17: 315–319.

MARSH, R.W. 1936. Notes on a technic for laboratory evaluation of protective fungicides. Br. Mycological Soc. Trans. 20: 304–309.

MASON, C., and D. POWELL. 1947. A Pythium plate method of evaluating fungicides. Phytopathology 37: 527–528.

MCCALLAN, S.E.A. 1930. Studies of fungicides 111. Testing protective fungicides in the laboratory. Cornell Agric. Exp. Stn. Memoir 128: 8–24.

MCCALLAN, S.E.A. 1947. Bioassay of agricultural fungicides. Agric. Chem. 2(9)31–34.

MCCALLAN, S.E.A. 1959. Problems and progress in plant pathology. 1908–1958. University of Wisconsin Press, Madison.

MCCALLAN, S.E.A., R.H. WELLMAN, and F. WILCOXON. 1941. An analysis of factors causing variation in spore germination tests of fungicides 111. Slope of toxicity curve, replicate tests, and fungi. Contrib. Boyce Thompson Inst. 12: 49–77.

MCCALLAN, S.E.A., and F. WILCOXON. 1940. An analysis of factors causing variation in spore germination tests of fungicides 11. Methods of spraying. Contrib. Boyce Thompson Inst. 11: 309–324.

MILLER, H.J. 1943. A comparison of laboratory and field retention and protective value of certain copper fungicides. Phytopathology 33: 899–909.

NEELY, D. 1969. The value of "in vitro" fungicide tests. Ill. Nat. Hist. Surv., Biol. Notes 64, p. 1–8.

PETERSON, P.D. 1941. The spore-germination method of evaluating fungicides. Phytopathology 31: 1108–1116.

SHARVELLE, E.G., and E.N. PELLETIER. 1956. A modified paper disk method of laboratory fungicide bioassay. Phytopathology 46: 26.

TORGESON, D.C. 1967. Determination and Measurement of Fungitoxicity Fungicides. An Advanced Treatise, Volume 1. Academic Press, New York.

TUITE, J.F. 1969. Plant Pathology Methods. Fungi and Bacteria. Burgess Publishing Co., Minneapolis, Minn.

WAGGONER, P.E., S. RICH, and J.G. HORSFALL. 1952. Prediction of field performance of fungicides. Phytopathology 42: 477.

WELLMAN, R.H., and F.D. HEALD. 1940. The toxicity of certain chemical aqueous solutions to spores of *Penicillium expansum*. Phytopathology 30: 638–648.

10

Compatibility of Fungicides

Plant diseases generally do not occur as sole predators of any particular crop, but are usually associated as partners in crime with insect pests, viruses, or other yield-reducing agents that prey upon the same host plant. Increasing crop production specialization and expansion of suburban living in recent years have focused attention on the need for controlling unwelcome pests in commercial agriculture and in the average homeowner's backyard garden. Chemicals for plant protection have so far survived attempts to purge the pesticides, and are recognized by many rational plant protectionists as necessary and indispensable tools for every farmer and backyard garden enthusiast.

The modern pesticides for plant disease, insect, and weed control are a far cry from the chemicals used prior to 1950. Today, we have effective, specific, safe, and relatively inexpensive compounds with long chemical names and complicated structural formulae to do the job of restricting the activities of crop saboteurs. So long as modern pesticides are used in the right way, at the right time, in proper amounts, and with recommended precautions, they will do the job for which they were developed. Failure to observe precautions and recommendations on the container label is usually the principal reason for the unsatisfactory performance of modern fungicides, insecticides, and herbicides.

For economy, the use of multiple combinations of pesticides is a necessary practice in commercial agriculture and is widely followed by most professional fruit, flower, and vegetable producers. The possible combinations of fungicides with other pesticides may be as follows:

a. Fungicides alone
b. Fungicides plus fungicides
c. Fungicides plus insecticides
d. Fungicides plus herbicides
e. Fungicides plus insecticides plus herbicides

f. Fungicides plus growth regulators
g. Fungicides plus insecticides plus growth regulators
h. Fungicides plus insecticides plus liquid nutrients plus minor elements

When pesticides are used in combination, or in mixtures, several problems may arise. In such cases the components of a pesticide mixture are said to be "compatible," or "incompatible."

COMPATIBILITY is the ability of two or more components of a pesticide mixture to be used in combination without impairment of toxicity, physical properties, or plant safety of either of the components. Thus, peaches and cream, beer and pretzels, corned beef and cabbage, and two Republicans in the same telephone booth are said to be "compatible."

INCOMPATIBILITY is a condition that may occur when two or more pesticides are used in combination, with resultant loss or impairment of effectiveness of either component: development of undesirable physical properties, reduction of toxicity, or the initiation of plant injury response (phytotoxicity). When any of these conditions occurs, the components of the combination are said to be "incompatible." Thus, ice cream and ketchup, stewed prunes and horseradish, kippers and custard, a coon hound and a bobcat in the back seat of a station wagon, and a Republican and a Democrat in the same voting booth are "incompatible." They just don't get along together.

Physical incompatibility is when two or more pesticides are mixed together and the result is an unstable mixture or a soapy flocculate. Thus, talcum powder is physically incompatible with water. Like oil and water, they do not mix together. Most organic fungicides are formulated with wetting agents or other conditioning materials. When they are forced through a sprayer screen with a high pressure stream of water, they occasionally separate out or "butter-out," and become physically incompatible. It is for this reason that most modern fungicides should be pre-pasted with a small quantity of water before being added to the sprayer tank.

Physical incompatibility frequently results when hard water from a deep well is used to mix the spray mixture. In such cases the fungicide will tend to fall out of suspension. This usually may be corrected by using pond soft water or adding commercial surfactants such as Santomerse, Igepon (sulfite lye), or Wetanol, or commercially prepared spray additives. When excessive foaming or frequent clogging of sprayer nozzles and screens occurs, or when heavy residues form on the sides of the sprayer tank, physical incompatibility, which should be investigated and corrected, is indicated.

Chemical incompatibility is when two or more pesticides are mixed together with resultant loss or reduction of effectiveness of one or all components. For this reason, most organic fungicides and insecticides should not be used in combination with alkaline compounds with a pH (hydrogen ion) concentration higher than pH 7.0. Alkaline reactions will usually significantly reduce the fungitoxicity of carbamate fungicides, as well as the insecticidal value of products such as Aramite, Lindane, Parathion, and Malathion. Lime applied for the prevention of moss in lawns should never be used in combination with maneb fungicides such as Fore, Manzate, or Dithane M-22. The same also applies to Dyrene, Captan, zineb, and Thiram.

Dinocap (Karathane), suggested for the prevention of powdery mildews, is chemically incompatible with calcium arsenate, oil-base formulations, and dinitro insecticides. Cycloheximide (Acti-Dione), suggested for the prevention of rust diseases of ryegrass, zoysia, bermudagrass, and St. Augustine grass, is chemically incompatible with chlordane and should never be used in combination with this insecticide. These are merely a few of the examples of chemical incompatibility of certain organic fungicides with some insecticides. Chemical incompatibility is frequently the cause of poor performance of some multiple pesticide combinations. Before combining any pesticide with another compound, it is important to read the directions on the label of the product package or container. If information on compatibility is not specified, it is always wise to avoid combination of pesticides of unknown chemical compatibility.

Phytotoxic incompatibility is when two or more pesticides used in combination result in injury to the host plant. They may be perfectly safe when used alone, but injurious in combination. For example, some organic fungicides may cause host injury when used in combination with emulsifiable concentrate (EC) insecticide formulations. This is especially true for those EC-insecticide formulations using xylene as the solvent for the toxic ingredient. When combining organic fungicides with liquid insecticides, check the label for compatibility and avoid plant injury problems. When combining pesticides of unknown compatibility, it is always a good suggestion to try the combination first on a non-essential or expendable plant or crop area before general use on a large scale.

Placement incompatibility is the incorrect placement of pesticides which may occasionally be the explanation for poor plant disease control. Placement incompatibility is less obvious than the problems previously discussed, and for this reason may often be overlooked. When two or more chemicals are used in combination and applied in one operation, each must end up in the right place if it is to do the job for which it was intended. Thus, most organic fungicides are "protectants" and must be uniformly distributed over susceptible plant surfaces of the host if they are to effectively protect against plant pathogens such as *Venturia inequalis* (apple scab) and *Alternaria solani* (tomato early blight). Failure to establish and maintain a fruit and foliar blanket of fungicide protection will inevitably result in outbreaks of destructive plant disease.

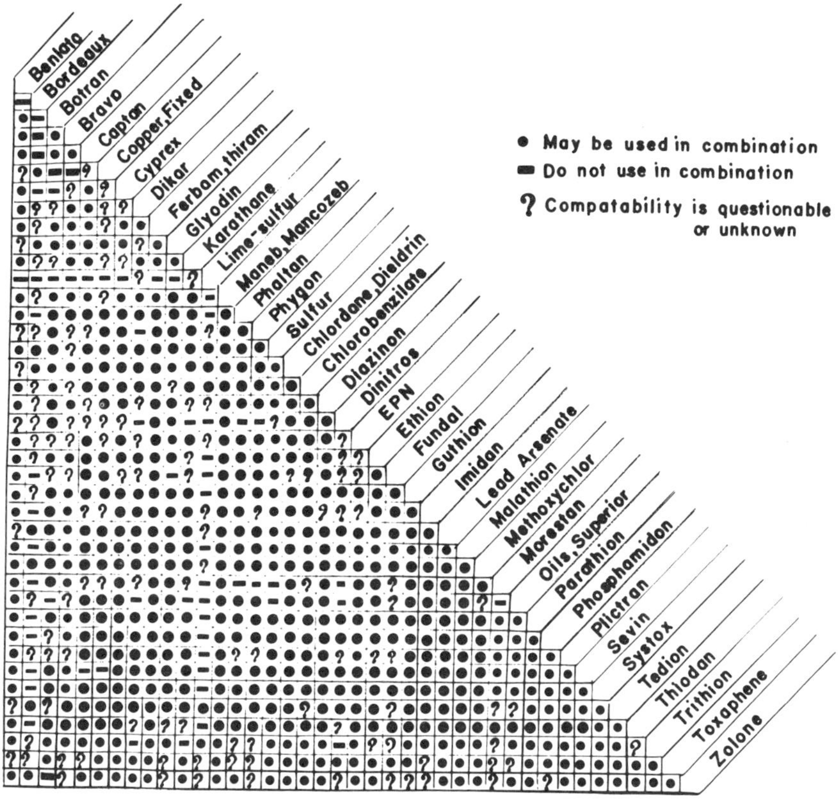

FIG. 10.1. COMPATIBILITY OF FUNGICIDES AND INSECTICIDES

In contrast to the above, copper fungicides for the control of mushroom fairy ring (*Agaricus campestris, Marasmius oreades,* etc.) in lawns are "eradicants" and should be applied to the soil surface, or within the upper soil levels, in order to eradicate the soil-borne fungi responsible for the fairy ring. Organic compounds used for the control of insects such as sod webworm, corn root worm, and cutworm would be ineffective if applied as foliar surface protectants, so must be placed where they will destroy the larval stages of the insect—namely the upper soil level. A combination of maneb fungicide for the control of gray leafspot disease of bermudagrass with the insecticide chlordane for sod webworm control would be ineffective because of placement incompatibility. Another example of the need for correct chemical placement is the use of ferrous sulfate, or iron chelate, for the correction of chlorosis (iron deficiency). To be effective, ferrous sulfate must be absorbed through the roots or by the leaves. Protective fungicides usually are not absorbed by plant leaf surfaces. Thus, if too much water is used when applying ferrous sulfate, some of it will be washed into the soil where it may become tied up and hence ineffective. Placement incompatibility is a very important consideration in combinations of some pesticides for crop protection.

The use of multiple pesticide combinations is a labor-saving short cut, but an understanding and knowledge of pesticide compatibility is essential in order to avoid problems which may arise from combination of some pesticides in "one-shot" applications. The details of specific compatibility of some fungicides with commonly used insecticides are summarized and illustrated in Fig. 10.1.

11

The Toxicity and Safe Use of Fungicides

With recent emphasis on contamination of the environment by the use of agricultural chemicals, the fungicides have been classified by the media, nature clubs, and many laymen as "hazardous chemicals" that should be eliminated from agricultural usage along with chlorinated hydrocarbons, mercury compounds, and other pesticides of proven mammalian toxicity. The scientific facts do not support the widely held opinion that chemicals for plant disease control—the fungicides—are all environmental contaminants which constitute a human hazard when used for plant disease prevention on edible food products. On the contrary, most of the modern fungicides have a relatively low order of mammalian toxicity and are subject to rigid restrictions to their use on edible food commodities.

The following discussion of the toxicity of modern fungicides is presented in an attempt to distinguish between fact and fiction, and to alleviate the apprehensions of those who are concerned that fungicides are a human hazard when edible commodities treated with these chemicals are consumed.

All fungicides are required by federal law, under the provisions of the Federal Insecticide, Fungicide, and Rodenticide Act (FIFRA), to be subjected to comprehensive mammalian toxicity tests by approved and qualified toxicologists before they can be labelled and distributed for use by commercial producers of edible food products.

The mammalian toxicity values for fungicides are based on laboratory tests with experimental animals (rats, white mice, and rabbits). Since toxicity tests with humans are illegal, the interpretation of fungicide toxicity values in terms of human toxicity and extrapolation of these values as to their effect upon humans should be done with discretion.

The toxicity values for fungicides are expressed as *Acute Toxicity, Chronic Toxicity, and Dermal Toxicity.*

Acute Toxicity is the amount of fungicide active ingredient administered in a single dose that is lethal to test animals (usually homozygous populations of white rats) under controlled laboratory conditions. Acute toxicity refers to *Acute Oral Toxicity* and *Acute Dermal Toxicity.*

Acute Oral LD_{50} is the dosage of active ingredient in a fungicide formulation administered orally in solution in one dose which will be lethal to 50% of the test population.

Acute Dermal LD_{50} toxicity is the dosage of active ingredient in a fungicide formulation which, when applied directly to the skin of test animals (usually rabbits), will be lethal to 50% of the test population. *Acute Oral and Dermal Toxicity values are expressed as milligrams per kilogram of body weight.*

Chronic Toxicity is the effect of long-term exposure of test animals to a fungicide and is determined by continually exposing the test animals to amounts of chemical less than the lethal amounts. Chronic Toxicity is a condition in which a chemical accumulates in the body, or its effects are additive, inducing illness or sometimes death. Chronic oral and Dermal toxicities are also expressed as milligrams per kilogram of body weight.

Acute Oral and Dermal toxicity values are the values most generally used for expressing the potential toxicity of fungicides. Mammalian toxicity values for fungicides are guides to the potential toxic effect of a compound to humans and animals. They do not constitute the only factor concerned with hazards associated with exposure to the compound. A compound may be highly toxic but present very little hazard when diluted and used according to the manufacturer's directions. Compounds of low mammalian toxicity can be hazardous when used in high concentrations or applied carelessly. Toxicity values of fungicides are used in part by the manufacturer to prescribe effective, safe dosages that will not violate established residue restrictions but will still control the plant diseases for which they were suggested.

On the basis of Acute Oral and Acute Dermal toxicity values determined by qualified and approved toxicologists, fungicides and other pesticides may be classified according to the Toxicity Classes summarized in Table 11.1.

On the basis of the data summarized in Table 11.2, it is evident that most of the fungicides used as foliage, fruit, and flower protectants are only slightly toxic and in many cases are characterized as practically non-toxic. Those fungicides that are classified as highly toxic are chemicals used for soil treatment or soil fumigation for the elimination of soil-borne plant pathogens. These fungicides include chloropicrin, methyl bromide, ethylene dibromide, and Dexon. They are not applied to edible plant products, require special equipment for their application, and,

TABLE 11.1. MAMMALIAN TOXICITY CLASSES FOR FUNGICIDES

Toxicity Rating	Commonly Used Term	Routes of Absorption		Probable Lethal Oral Dose for 150-lb Man
		LD_{50} Single Oral Dose Rats mg/kg	LD_{50} Single Dermal Dose Rabbits mg/kg	
1	Extremely Toxic	1 mg or less	20 mg or less	A taste, a grain
2	Highly Toxic	1–50 mg	20–200 mg	A pinch, 1 tsp
3	Moderately Toxic	50–500 mg	200–1,000 mg	1 tsp – 2 tbsp
4	Slightly Toxic	500–5,000 mg	1,000–2,000 mg	1 oz – 1 pt
5	Practically Non-toxic	5,000–15,000 mg	2,000–20,000 mg	1 pt–1 qt
6	Relatively Harmless	Greater than 15,000 mg	Greater than 20,000 mg	Greater than 1 qt

Source: Adapted from Gleason, Gosselin, and Hodges (1963) and Hayes (1963).

TABLE 11.2. ACUTE ORAL AND DERMAL MAMMALIAN TOXICITY VALUES FOR SOME MODERN FUNGICIDES (MILLIGRAMS/KILOGRAM OF BODY WEIGHT)

Name of Fungicide	Oral LD_{50} (mg/kg)	Dermal LD_{50} (mg/kg)	Toxicity Class (on Basis of Oral Toxicity)
Acti-dione	*See* Cycloheximide		
Agrox–M	*See* Maneb		
anilazine	*See* Dyrene		
Arasan	*See* Thiram		
Aspirin	750	3,830	3 (MT)
Benlate	*See* benomyl		
benomyl	>10,000	L	5 (NT)
binapacryl (Morocide)	421	—	4 (ST)
Bordeaux mixture	Non-Toxic	> 1,000	3 (MT)
Botran	*See* dicloran		
Bravo	*See* daconil		
Brom-O-Gas	*See* Methyl Bromide		
Calcium polysulfide (Lime-sulfur)	Low	M	6 (H)
captafol (Difolatan)	6,200	S	5 (NT)
Captan	10,000	L	5 (NT)
carboxin (Vitavax)	3,820	> 8,000	4 (ST)
chloroneb (Demosan)	>11,000	> 3,000	5 (NT)
Chloropicrin	250	S	1 (ET)
chlorothalonil (Bravo) (Daconil)	>10,000	>10,000	5 (NT)
Copper hydroxide	1,000	S	4 (ST)
cyloheximide (Acti-dione)	133	S	2 (HT)
Cyprex	*See* dodine		
Daconil	*See* chlorothalonil		
D-D Mixture	140	2,100	3 (MT)
Demosan	*See* chloroneb		
Dexon	*See* fenominosulf		
dichlone (Phygon)	1,300	S	4 (ST)
dichloropropene (Telone)	250	S	3 (MT)
Difolatan	*See* captafol		
Dikar	*See* zinc-ion maneb or dinocap		
dinocap (Karathane)	980	9,400	4 (ST)
Dithane D-14	*See* nabam		
Dithane M -22	*See* maneb		
Dithane M -45	*See* zinc-ion maneb		
Dithane Z - 78	*See* zineb		
DNOC (Elgetol)	30	M	2 (HT)
dodine (Cyprex)	1,000	> 1,500	4 (ST)
Dowfume MC-2	*See* methyl bromide		
Du-Ter	*See* Fentin hydroxide		
Dyrene	*See* anilazine		
Elgetol	*See* DNOC		
fenaminosulf (Dexon)	64	100	3 (MT)
fentin hydroxide (Du-Ter)	108	5,000	3 (MT)
ferbam (Fermate)	>17,000	> 1,000	6 (RH)
Ferradow	*See* ferbam		
folpet (Phaltan)	>10,000	M	5 (NT)
Fore	*See* zinc-ion maneb		
formaldehyde (Formalin)	800	M	4 (ST)
Glyodin	4,600-7,600	M	4 (ST)
Karathane	*See* dinocap		

TABLE 11.2 *(Continued)*

Name of Fungicide	Oral LD_{50} mg/kg	Dermal LD_{50} mg/kg	Toxicity Class (On Basis of Oral Toxicity)
Karbam	*See* ferbam		
Kocide	*See* copper hydroxide		
Lime	*See* calcium hydroxide		
Lime-sulfur	*See* calcium polysulfide		
maneb	6,750	7,100	5 (NT)
Manzate	*See* maneb		
Manzate 'D'	*See* maneb		
Manzate 200	*See* zinc-ion maneb		
Mertect	*See* thiabendazole		
Methyl bromide	1 mg/liter	S	1 (HT)
Mildex	*See* dinocap		
Morestan	2,500	> 2,000	4 (ST)
Morocide	*See* binapacryl		
nabam (Parzate)	395	> 1,000	3 (MT)
Niacide	*See* ferbam		
Orthocide	*See* Captan		
Ortho zineb	*See* zineb		
oxycarboxin (Plantvax)	2,000	>16,000	4 (ST)
Panoram	*See* Thiram		
parinol (Parnon)	5,000	—	4 (ST)
PCNB	> 1,700	M	5 (NT)
Phaltan	*See* folpet		
Phygon	*See* dichlone		
piperalin	2,500	?	4 (ST)
Pipron	*See* piperalin		
Plantvax	*See* oxycarboxin		
Polyram	6,200	> 2,000	4 (ST)
Streptomycin	9,000	> 600	5 (NT)
Sulfur	>17,000	M	6 (H)
Table salt	3,320	—	4 (ST)
Telone	*See* dichloropropene		
Termil	*See* chlorothalonil		
Terraclor	*See* PCNB		
Terrazole	1,077	?	4 (ST)
Tersan	*See* Thiram		
Thiram (Thylate)	780	S	4 (ST)
Vapam	820	4,640	4 (ST)
Vitavax	*See* oxycarboxin		
Vorlex	305	M	3 (MT)
Zinc-ion maneb	> 8,000	> 8,000	5 (NT)
zineb	> 5,200	> 1,000	5 (NT)
ziram	1,400	> 1,000	4 (ST)

Source: Anon. (1978).
\> = greater than.
ET = extremely toxic.
HT = highly toxic.
MT = moderately toxic.
ST = slightly toxic.
NT = practically non-toxic.
H = harmless.
L = low dermal toxicity.
M = moderate dermal toxicity.
S = severe dermal toxicity.

being highly volatile and biodegradable, do not persist in treated soil. Other highly toxic products such as cycloheximide are used only in extreme dilutions which are not toxic to the operator or to edible products to which they are applied. Acti-dione, recommended for the control of powdery mildew, contains only 0.027% active ingredient, and a dosage of 1-2 level tablespoonfuls per gallon of water is suggested.

A comparison of the relative mammalian toxicities of various protective fungicides as compared with some insecticides and common domestic products is illustrated in Fig. 11.1. The modern fungicides currently recommended for the control of common diseases of fruits, vegetables, turf, and ornamentals are among the safest and least toxic of agricultural pesticides.

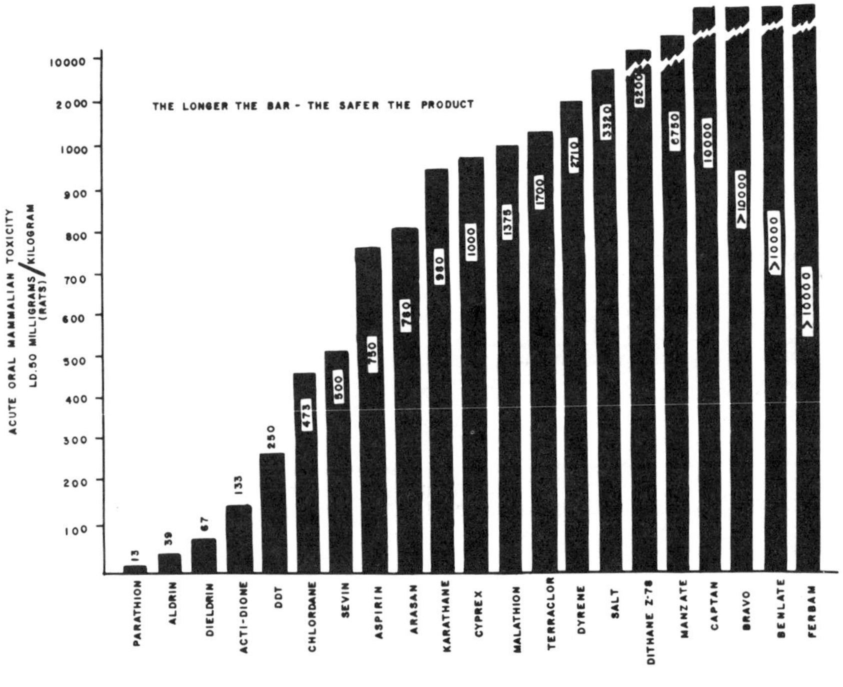

FIG. 11.1. A COMPARISON OF THE ACUTE ORAL MAMMALIAN TOXICITY OF COMMON FUNGICIDES WITH INSECTICIDES AND SOME HOUSEHOLD PRODUCTS

FUNGICIDES APPROVED FOR USE ON VEGETABLES AND FRUIT CROPS

Under the provisions of FIFRA (Federal Insecticide, Fungicide, and Rodenticide Act), it is illegal to use any pesticide on any edible crop unless that product has been approved for use on the commodity by federal pesticide regulatory agencies. The Federal Pesticide Control Act of 1972 declares it to be a federal offense if any pesticide is used by a grower in a manner inconsistent with the label on the specific package of pesticide being used. Although approved uses of pesticides are likely to change at any time, partial listings of approved fungicides for vegetables and fruit crops are summarized in Table 11.3 and Table 11.4.

TABLE 11.3. APPROVED FUNGICIDES FOR USE ON VEGETABLES (AS OF MAY, 1979)

Vegetable	Fungicide	Vegetable	Fungicide
Asparagus	Bordeaux mixture	Carrots	Botran
	Captan		Bravo
	Fixed copper		Captan
	Maneb		Fixed coppers
	Polyram		Maneb
	Zinc-ion maneb		Thiram
	Zineb		Zinc-ion maneb
	Vorlex		Zineb
	Vapam		DD
	DD		DCP
	DCP		Vorlex
			Vapam
Beets	Bordeaux mixture		
	Captan	Collards	Captan
	Fixed coppers		Maneb
	Thiram		Thiram
	Zineb		Zineb
	Vorlex		DD
	Vapam		DCP
	DD		Vorlex
	DCP		Vapam
Cabbage	Bravo	Cucumber	Benlate
	Captan		Botran
	Fixed coppers		Bravo
	Maneb		Captan
	Terraclor		Difolatan
	Thiram		Dyrene
	Zineb		Fixed coppers
	Vorlex		Karathane
	Vapam		Maneb
	DD		Polyram
	DCP		Thiram
			Zinc-ion maneb
Canteloupe	Benlate		Zineb
	Bravo		DD
	Captan		DCP
	Difolatan		Vorlex
	Dyrene		Vapam
	Fixed coppers		
	Karathane	Eggplant	Captan
	Maneb		Fixed coppers
	Polyram		Maneb
	Thiram		Thiram
	Zinc-ion maneb		Zineb
	Zineb		DD
	Vorlex		DCP
	Vapam		Vorlex
	DD		Vapam
	DCP		

TABLE 11.3 *(Continued)*

Vegetable	Fungicide	Vegetable	Fungicide
Endive	Maneb	Potatoes	Botran
	Zineb		Bravo
	DD		Captan
	DCP		Difolatan
	Vorlex		Du-Ter
	Vapam		Dyrene
			Fixed coppers
Lettuce	Botran		Maneb
	Captan		Polyram
	Fixed coppers		Terraclor
	Maneb		Zinc-ion maneb
	Thiram		Zineb
	Zineb		DD
	DD		DCP
	DCP		Vapam
	Vapam		Vorlex
	Vorlex		
		Pumpkin	Bravo
Mustard	Captan		Captan
	Maneb		Dyrene
	Thiram		Fixed coppers
	Zineb		Karathane
	DD		Maneb
	DCP		Thiram
	Vapam		Zineb
	Vorlex		DD
			DCP
Onions	Botran		Vapam
	Bravo		Vorlex
	Captan		
	Difolatan	Radishes	Thiram
	Dyrene		Zineb
	Fixed coppers		DD
	Maneb		DCP
	Thiram		Vapam
	Zinc-ion maneb		Vorlex
	Zineb		
	DD	Rhubarb	Botran
	DCP		Captan
	Vapam		Maneb
	Vorlex		DD
			DCP
Parsley	Zineb		Vapam
	DD		Vorlex
	DCP		
	Vapam	Snap Beans	Benlate
	Vorlex		Botran
			Bravo
Peppers	Captan		Captan
	Fixed coppers		Fixed coppers
	Maneb		Maneb
	Terraclor		Terraclor
	Zineb		Thiram
	DD		Zineb
	DCP		DD
	Vapam		DCP
	Vorlex		Vapam
			Vorlex

TABLE 11.3 *(Continued)*

Vegetable	Fungicide
Spinach	Captan Dyrene Fixed coppers Maneb Thiram Zineb DD DCP Vapam Vorlex
Squash	Benlate Bravo Captan Fixed coppers Karathane Maneb Thiram Zinc-ion maneb Zineb DD DCP Vapam Vorlex
Sweet Corn	Bravo Captan Maneb Polyram Thiram Zinc-ion maneb Zineb DD DCP Vapam Vorlex
Sweet Potato	Botran Thiram DD DCP Vapam Vorlex
Tomatoes	Benlate Botran Bravo Captan Difolatan Dyrene Fixed coppers Maneb Polyram Terraclor Thiram Zinc-ion maneb Zineb DD DCP Vapam Vorlex
Turnips	Captan Maneb Thiram Zineb DD DCP Vapam Vorlex
Watermelons	Benlate Bravo Captan Difolatan Dyrene Fixed coppers Karathane Maneb Thiram Zinc-ion maneb Zineb DD DCP Vapam Vorlex

BEFORE USING ANY OF THE FUNGICIDES LISTED, BE SURE TO READ THE LABEL ON THE PACKAGE OR CONTAINER. FOLLOW ALL DIRECTIONS CAREFULLY. TO USE A FUNGICIDE IN A MANNER INCONSISTENT WITH THE LABEL IS A VIOLATION OF FEDERAL LAW.

TABLE 11.4. APPROVED FUNGICIDES FOR USE ON FRUIT CROPS (AS OF MAY, 1978)

Fruit	Fungicide
Apples	Benlate
	Bordeaux mixture
	Captan
	Difolatan (dormant only)
	Phygon
	Dikar
	Karathane
	DNOC (dormant only)
	Cyprex
	Ferbam
	Phaltan
	Glyodin
	Lime-sulfur
	Mancozeb
	Niacide 'M'
	Polyram
	Streptomycin
	Sulfur
	Thiram
	Zineb
Blackberries	Benlate
	Bordeaux mixture
	Captan
	Ferbam
	Phaltan
	Lime-sulfur
	Sulfur
	Zineb
Blueberries	Benlate
	Captan
	Ferbam
	Phaltan
	Sulfur
Cherries	Benlate
	Bordeaux mixture
	Botran
	Captan
	Difolatan**
	Phygon
	Cyprex
	Ferbam
	Phaltan
	Glyodin
	Lime-sulfur
	Sulfur
	Zineb
Grapes	Benlate
	Bordeaux mixture
	Botran
	Captan
	Karathane
	Ferbam
	Phaltan
	Lime-sulfur
	Mancozeb
	Sulfur
	Zineb
Peaches	Benlate
	Bordeaux mixture
	Botran
	Captan
	Difolatan**
	Phygon
	Karathane
	Cyprex (dormant only)
	Ferbam
	Glyodin
	Lime-sulfur
	Sulfur
	Thiram
	Zineb
Pears	Benlate
	Bordeaux mixture
	Captan
	Karathane
	Cyprex
	Ferbam
	Glyodin
	Lime-sulfur
	Mancozeb
	Streptomycin
	Sulfur
	Zineb
Plums and Prunes	Benlate
	Bordeaux mixture
	Botran
	Captan
	Difolatan**
	Phygon
	Ferbam
	Lime-sulfur
	Sulfur
	Zineb
Raspberries	Benlate
	Bordeaux mixture
	Captan
	Ferbam
	Phaltan
	Lime-sulfur
	Sulfur
	Zineb
Strawberries	Benlate
	Bordeaux mixture

TABLE 11.4 *(Continued)*

Fruit	Fungicide	Fruit	Fungicide
		Strawberries (Cont.)	Captan Cyprex Ferbam Phaltan Phygon Sulfur Thiram Zineb

** Refer to restrictions on label.

FUNGICIDE TOLERANCES

Under the terms of the Federal Insecticide, Fungicide, and Rodenticide Act (FIFRA), as administered by the Environmental Protection Agency (EPA), the residues of various pesticides, including fungicides, that are permissible on an edible agricultural product at harvest time are also regulated.

A *pesticide residue* may be defined as "the remnant of a pesticide, or its degredation products, that can be found on or in a crop or commodity after application or other exposure to a pesticidal chemical. Residues may result from a direct application, drift, up-take from contaminated soil, or from other environmental sources. Residues are usually expressed as parts per million (PPM) on a weight basis."

One part per million is one ounce of pesticide residue in or on one million ounces of the agricultural product. The equivalent of 1.0 ppm is 1 oz of sand in 31.25 tons of cement, or 1 oz of salt in 62,500 lb of sugar.

Since the provisions of FIFRA require the pesticide manufacturer to develop analytical techniques and analyses, and since the EPA administers the residue tolerances established by the Federal Food and Drug Administration (FDA), it is possible to detect amounts and types of many pesticides at levels approaching 0.001 ppm or one part per billion.

A *residue tolerance* is the amount of a pesticide which qualified toxicologists have determined may safely remain on or in the food crop without hazard to the ultimate consumer. Permissible residues at harvest time for fungicides on or in fruits and vegetables have been established by the FDA. These permissible residues are expressed as parts per million of the active ingredient of the product on the commodity at harvest time. Specific permissible residues for fungicides and other pesticides are listed in publications of the FDA. However, residue tolerances have little practical function for the commercial or amateur users of fungicides since laboratory equipment and analytical procedures are impractical or unavailable to most users of fungicides for plant disease control.

To help fungicide users avoid excesses of established harvest residues, the system of "Cut-off Dates," or discontinuation of fungicide usage at specified days before harvest, has been suggested as a means of avoiding crop residues of fungicides in excess of legal limits.

The "cut-off dates" for fruit fungicides are summarized in Table 11.5, and in Table 6.3 for vegetable fungicides.

THE TWELVE GOLDEN RULES OF FUNGICIDE SAFETY

Although the modern fungicides currently suggested for plant disease control are relatively non-toxic, certain precautions should be followed when they are used. These safety precautions are summarized below.

1. Identify the plant disease problem and use the fungicide recommended and approved for its control.
2. Store fungicides in locked cabinets or other receptacles in the garage or tool shed, out of reach of children and domestic animals, and labelled with the name of the fungicide.
3. Fungicides should be weighed out and prepared for application in an open area where ventilation is adequate, and should never be prepared in an enclosed, unventilated area.
4. Never spray or dust with fungicides on a windy day, and stay out of the spray or dust drift.
5. Do not smoke when applying fungicides.
6. Wash hands, face, arms, and other exposed body areas with soap and water after applying fungicides.
7. Change contaminated clothing after applying fungicides.
8. Clean fungicide sprayers or dusters thoroughly after use.
9. Never use spray applicators that previously have been used for herbicide applications without careful decontamination. Preferably reserve sprayers and dusters for fungicides only.
10. Use fungicides in the amounts recommended by the manufacturer and weigh out the recommended dosages accurately. Two tablespoonfuls per gallon of water means TWO TABLESPOONFULS. Never add "just a little bit more to make sure it controls the problem."
11. Dispose of empty paper or cardboard fungicide containers by burning and dispose of glass and metal containers in a safe place.
12. When in doubt, seek the advice of qualified plant pathologists, extension specialists, or the county agricultural agent before making mistakes.

THERE IS A SAFE SIDE TO EVERY PESTICIDE. DON'T GAMBLE WITH FUNGICIDES. FOLLOW THE DIRECTIONS ON THE CONTAINER.

TABLE 11.5. NUMBER OF DAYS FROM LAST SPRAY TO HARVEST (CUT-OFF DATES) FOR FRUIT FUNGICIDES

Fungicide	Apples	Apricots	Blackberries	Blueberries	Cherries	Grapes	Peaches	Pears	Plums and Prunes	Raspberries	Strawberries
Benlate (Benomyl)	7	0	3	21	0	7	0	7	0	3	0
Bordeaux mixture	0	0	0	0	0	0	0	0	0	0	0
Botran (dicloran)	—	—	—	—	1	1	1	—	PH PB	—	—
Captan	0	0	0	0	0	0	1	0	0	0	0
Difolatan (captafol)	*	—	—	21	*	—	*	—	*	—	—
Phygon (dichlone)	1	—	—	—	3	—	7	—	3	—	3
Dikar	21	—	—	—	—	—	—	—	—	—	—
Karathane (dinocap)	21	—	—	—	—	21	45	21	—	7	21
Cyprex (dodine)	7	—	—	—	0	—	15	7	—	—	14
Fermate (ferbam)	7	—	—	40	0	7	21	7	7	40	14
Phaltan (folpet)	0	—	—	0	0	0	—	—	—	0	0
Glyodin	0	—	—	—	7	—	0	0	—	—	—
Lime-sulfur	0	—	—	—	0	—	0	0	0	—	—
Zinc-ion maneb	21	—	—	—	0	—	0	0	0	—	—
Polyram	30	—	—	—	—	—	—	—	—	—	—
Streptomycin	50	—	—	—	—	—	—	30	—	—	—
Sulfur	0	—	—	0	0	0	0	0	0	—	—
Thiram (Thylate)	0	—	—	—	—	—	7	—	—	—	—
Zineb	30	—	—	—	7	7	30	14	30	—	—

PB = Pre-bloom only.
PH = Postharvest only.
* Refer to label restrictions.
– = Not registered.

BIBLIOGRAPHY

ANON. 1976A. Safe, effective use of pesticides. A manual for private applicators. Michigan State University. Coop. Ext. Ser. Ext. Bull. *E 1025.* Michigan State Univ., East Lansing.

ANON. 1976B. Safe, effective use of pesticides. A manual for commercial applicators. Michigan State University. Coop. Ext. Ser. Ext. Bull. *E 1032.* Michigan State Univ., East Lansing.

ANON. 1978. Farm Chemicals Handbook. Meister Publications, Willoughby, Ohio.

BAILEY, J.B., and J.E. SWIFT. 1976. Pesticides Information and Safety Manual. Univ. California, Div. Agric. Sci. Sale Publ. 4037. Univ. of California, Berkeley.

BRAZELTON, R.W., N.B. AKESSON, and W.E. YATES. 1976. Study Guide for Pest Advisors on the Safe Application of Agricultural Chemicals. Univ. California, Div. Agric. Sci. Sale Publ. 4048. Univ. of California, Berkeley.

BROGDON, J.E. *et al.* 1976. Apply pesticides properly. University of Florida Center for Environmental Programs. Institute of Food and Agricultural Sciences, Gainesville.

DEWEY, J.E., S.A. MACDONALD, R.F. PENDLETON, and W.G. SMITH. 1974. Pesticide Applicator Training Manual. Core Manual. Northeastern Pesticide Coordinators Coop. Ext. Ser., Cornell University, Ithaca, N.Y.

GLEASON, M.N., R.E. GOSSELIN, and H.C. HODGE. 1963. Clinical Toxicology of Commercial Products. 2nd edition. Williams and Wilkins Co., Baltimore.

HAYES, W.J. 1963. Clinical Handbook of Economic Poisons. U.S. Dep. Health, Education & Welfare. Public Health Ser. Communicable Disease Center. Toxicology Section. Atlanta.

APPENDIX

A. ABBREVIATIONS USED IN PLANT DISEASE CONTROL

A	= Acre(s)	gm	= Gram(s)
ae	= Acid equivalent	gal.	= Gallon(s)
aehg	= Acid equivalent per 100 gallons	gpa	= Gallons per acre
		gpm	= Gallons per minute
ai	= Active ingredient(s)	gr	= Grain(s)
aihg	= Active ingredient per 100 gallons	hr	= Hour(s)
		ht	= Height
approx	= Approximately	in.	= Inch(s)
Avdp	= Avoirdupois	Kg	= Kilogram(s)
BP	= Boiling Point	L	= Liter(s)
bu	= Bushel(s)	lb	= Pound(s)
c	= Cup(s)	mg	= Milligram(s)
C	= Centigrade	mcg	= Microgram(s)
cc	= Cubic centimeter(s); same as milliliters	mi	= Mile(s)
		min	= Minute(s)
cf	= Cubic feet	ml	= Milliliter; same as cc
cm	= Centimeter(s)		
C.P.	= Chemically Pure	mm	= Millimeter(s)
cwt	= Hundredweight(s)	MP	= Melting Point
cu	= Cubic	mph	= Miles per Hour
diam	= Diameter	MW	= Molecular Weight
EC	= Emulsifiable Concentrate	oz	= Ounce(s)
F	= Fahrenheit	ppm	= Parts per million
fl	= Fluid	ppmv	= Parts per million by volume
fpm	= Feet per minute		
ft	= Foot or feet	ppmw	= Parts per million by weight

ppt	=	Precipitate
psi	=	Pounds per square inch
pt	=	Pint(s)
Q.S.ad	=	Quantity sufficient to make
qt	=	Quart
rd	=	Rod(s)
rpm	=	Revolutions per minute
rps	=	Revolutions per second
sat	=	Saturated
sol	=	Solution(s)
sp.gr.	=	Specific gravity
sq	=	Square
T	=	Ton(s)
tbsp	=	Tablespoon(s)
temp	=	Temperature
tsp	=	Teaspoon(s)
WP	=	Wettable Powder
wt	=	Weight
μ	=	Micron
mμ	=	Millimicron
%	=	Percent
<	=	Less Than
>	=	More Than
--	=	Equal Parts
aa	=	Degree

B. A DICTIONARY OF TERMINOLOGY USED FOR PLANT DISEASES AND THEIR CONTROL

The definitions summarized alphabetically herewith are based on the terminology used in standard Phytopathological references, and on the definitions adopted by the American Association of Pest Control Officials, Inc.

ABSORPTION	Movement of a pesticidal chemical from the surface into a plant, animal, or the soil
ACID	*See* pH
ACID EQUIVALENT	That portion of a compound or formulation that theoretically could be converted back to the corresponding acid
ACRE	An area of land measuring 210 ft long and 210 ft wide, containing 43,560 sq ft equal to 0.4047 hectares
ACTIVATOR	A material added to a pesticide to increase its toxicity either directly or indirectly
ACTIVE INGREDIENT	A substance contained in a preparation which will by itself act in the same manner and for the same purposes as the directions provided for the preparation of the whole. It must be present in an amount sufficient to add materially to the effectiveness of the preparation
ACTUAL DOSAGE	The amount of a pure chemical in a pesticide formulation used per unit area, unit volume, or per individual plant, e.g., 2.0 lb of 50% WP Captan in 100 gal. of water to one acre is 1.0 lb per acre actual dosage

ACUTE ORAL LD_{50}	The single dosage of a chemical fed orally to a test animal that will kill 50% of the test population
ACUTE POISONING	Illness resulting from a single dosage or dosages administered over a short period of time
ACUTE TOXICITY LD_{50}	The total dosage of a pesticide fed over a period of time to test animals that will kill 50% of the test population
ADHERENCE	The property of a fungicide or other pesticide to adhere to, or to stick to, a specified surface
ADJUVANT	Synonymous with "additive." An ingredient which, when added to a formulation, aids in the action of the toxicant. The term includes such materials as wetting agents, spreaders, adhesives, emulsifiers, dispersing agents, penetrants, and correctives
ADSORPTION	Condensation or adhesion of gases and liquids on the surfaces of solids
ADVANCING CONTACT ANGLE	The angle formed between the spray deposit on the host or other surface and the surface of the host or other surface
AEROBIC	Capable of living or functioning in air or free oxygen
AEROSOL	A system of particles dispersed in a gas. Liquid particles make up a fog and solid particles form a smoke. The term usually refers to liquids in the case of pesticide aerosols

AIR-BLAST APPLICATIONS	The application of a chemical to a surface or area using high velocity air from specialized application equipment
ALGICIDE	A chemical intended for the control of algae, especially in water that is stored or is being used industrially
ALKALINE	*See* pH
ALKALOID	A physiologically active, usually naturally occurring, nitrogenous compound that is alkaline in reaction. Many are characteristic of certain plants, e.g., nicotine in tobacco
ALLOTROPIC	Originating from different areas or regions
ALTERNATE HOST	One or other of the two unlike hosts of a heteroecious rust
AMPHOTERIC	An amphoteric compound has the capacity of behaving either as an acid or a base. An amphoteric surface-active compound is capable of anionic or cationic behaviour, depending on whether it is an acid or basic system
ANAEROBIC	Living or functioning in the absence of air or free oxygen
ANIONIC	An ion having a negative charge is an "anion." When the surface-active portion of a surfactant molecule possesses a negative charge it is termed "an anionic surface-active agent"
ANNUAL	A plant which completes its life cycle from seed to maturity in one year

ANTAGONISM	The phenomenon which results when two or more organisms occur in close proximity, resulting in depression of the growth of one of the organisms or in a reduction of activity of a chemical compound
ANTIBIOTIC	Any of certain chemical substances produced by microorganisms such as bacteria and fungi (molds), and having the capacity to inhibit the growth or to destroy bacteria and certain fungi causing diseases of animals and plants, e.g., Aureomycin, Streptomycin, Terramycin, Phytoactin
ANTIDOTE	An immediate or first aid treatment or remedy to offset the harmful effects of a pesticidal chemical or other toxicant
ANTIOXIDANT	A substance capable of chemically protecting other substances against oxidation or spoilage
APPLICATION RATE	Similar to dosage but more inclusive, e.g., an application of 50% WP Captan at a dosage of 2 lb per acre in 100 gal. of water
ATTRACTANT	A chemical or agent that attracts or lures other life, e.g., light is an attractant for moths
ATTRAHANT	Same as "attractant"
AUTOECIOUS	A condition whereby a parasitic fungus is able to complete its life cycle on one host. Generally applied to rust fungi
AUXILIARY MATERIAL	An agency added to a pesticide to improve its physical properties

AUXIN	A generic term for compounds characterized by their capacity to elongate plant cells and resembling indole-3-acetic acid in their physiological activity
AVERAGE PARTICLE SIZE	Since the particles of most fungicides are frequently non-spherical, it is difficult to define the parameters used for particle size. The parameters most frequently used are the maximum chord, the projected area, the distance between two parallel tangents, and the maximum chord in a given direction
AVIRULENT	Non-pathogenic
BACTERICIDE	A pesticidal chemical or other agent which inhibits or prevents the growth and development of bacteria
BACTERIOSTAT	A material that prevents growth or multiplication of bacteria without being toxic
BAND APPLICATION	An application to a continuously retricted area such as in or along a crop row rather than over the entire area
BASAL TREATMENT	A treatment applied to the stems or trunks of plants at or immediately above the soil line
BED	(a) A narrow flat-topped ridge on which crops are grown; (b) an area in which seedling plants are produced for transplanting
BENTONITE	A colloidal native clay (hydrated aluminum silicate) which has the property of forming viscous suspensions or gels when mixed with water. Used in

	dust formulations to improve tenacity and weathering
BIOASSAY	A method for measuring the amount of toxicant in a test material by the reaction of a test animal or test plant to the product. Fungicidal bioassay refers to laboratory, greenhouse, or field evaluation of the fungicidal potential of a test chemical
BIOCIDE	A compound that is toxic to all forms of life
BIOLOGICAL CONTROL	Parasitic and predacious insects and insect disease organisms are vital in the natural control of insects and other pests. Such forms may also often be raised or reared and disseminated artificially. Biological control also includes the use of insects to control weeds as well as the use of other living organisms in fighting pests
BLIGHT	A general term that may include spotting, discoloration, sudden wilting, or death of leaves, flowers, fruits, stems, or the entire plant
BLOTCH	A leaf symptom involving formation of large areas of dead leaf tissue. May also refer to affected areas on fruits
BRAND NAME	Same as "trade name." Refers to the name, number, trademark, or designation applied to an economic poison of any particular description by the manufacturer, distributor, importer, or vendor thereof. Each economic poison differing in the ingredient statement analysis, manufacturer or distributor name, number, or trademark shall be considered as a separate and distinct brand

BROADCAST APPLICATION	An application over an entire area
BROAD SPECTRUM FUNGICIDE	A fungicide that controls a wide range of pathogens causing plant diseases
CALLUS	An overgrowth produced by a plant in response to wounding by fungi, insects or mechanical damage
CANKER	A lesion on a plant stem or twig, often swollen or sunken, surrounded by living tissue
CARCINOGEN	A substance that may cause cancer in experimental animals, or is known to do so in humans
CARDINAL TEMPERATURES	The minimum, optimum, and maximum temperatures for fungus growth and development
CARRIER	An inert material added to a technical poison or other material, to facilitate later dilution to field strength in simple mixing equipment. Solid carriers are used in the preparation of wettable powders, dust concentrates, and granules, and must be compatible with the toxicant, having good absorption or adsorption properties, be free-flowing, non-abrasive, of fine particle size, and for wettable powders must be easily dispersed in water with high suspension value. Although generally considered inert they have a definite effect on potency and efficiency of dusts and sprays since most formulations contain from 80–90% carrier-diluent
CATALYST	A substance that accelerates the rate of a chemical reaction but is not itself used up in the reaction

CATIONIC	An ion having a positive charge is a cation. When the surface-active portion of a surfactant molecule possesses a positive charge, it is termed a "cationic surfactant"
CAUSAL AGENT	An agent such as a fungus, bacterium, virus, nematode, etc., that incites a given plant disease
CENTIGRADE	A temperature scale on which water freezes at 0°C and boils at 100°C
CHEMICAL NAME	The name given to a pesticide based on its chemical composition, e.g., Maneb is a formulation containing manganese ethylene bisdithiocarbamate
CHEMOSTERILANT	A pesticidal chemical which destroys the ability of microorganisms and pests to reproduce
CHEMOTHERAPY	The treatment of a plant disease with a chemical to destroy or inactivate a pathogen without affecting the host plant
CHLOROSIS	Yellowing of normally green tissue due to destruction of chlorophyll or partial failure of chlorophyll to develop
CHRONIC DISEASE	Lasting a long time; a disease of long duration; opposed to acute
CHRONIC TOXICITY	Effect of long term exposure to a pesticidal chemical by a test organism. Test results are obtained by continuously exposing animals to amounts of the chemical less than the amount required to kill them in one dose

CIRCULATIVE — Characterizes those plant viruses that are transmitted by the mouth parts of the insect vector, accumulated internally, passed through the insect tissues, and then transmitted by the vector's mouth parts

CLAY — A disperse system of mineral fragments of hydrated aluminum silicate, predominantly those smaller than 2.0 microns, which are plastic when wet and permanently hard when "fired." The term clay refers to a physical condition, not a definite chemical composition. The four basic types of clay are kaolite, illite, attapulgite, and montmorillonite. They are formed as weathering products from rocks and minerals. The four types are vastly different from one another

COINED NAME — *See* "common name"

COMMON NAME — The name given to a chemical by a recognized agency such as the American Standards Committee on Common Names, the USDA, and the American Phytopathological Society, e.g., the common name of zinc ethylenebisdithiocarbamate is "zineb"

COMPATIBILITY — The ability of two or more substances to mix without objectionable change in their physical or chemical properties

CONCENTRATE — A liquid or dry formulation containing a percentage of toxicant high enough to save shipping charges and yet be of convenient strength for easy dilution

CONCENTRATION	The percentage or the amount by weight of a pesticidal chemical in a formulation or preparation ready for application
CONTAMINATE	To make feed, food, water, soil, clothing, etc., unfit for specified use by accidental or careless use, handling, or disposal of a pesticidal chemical
COVERAGE	The degree or amount of spread of a pesticidal chemical over a given surface
CROP TOLERANCE	The degree or ability of a crop to be treated with a chemical without adverse effect, OR the legal permissible residue of a pesticidal chemical at harvest time permitted by official regulatory authority
DAMPING-OFF	Decay of seeds in the soil, or of young seedlings before or after emergence. Generally caused by seed-borne or soil-borne fungi
DEFLOCCULATOR	A dispersing agent used to retard settling of solid particles in suspension, especially when the particles tend to clump together and settle out rapidly. Emulsifiers are often effective deflocculators
DEFOLIANT OR DEFOLIATOR	A chemical that causes the leaves of a plant treated with the compound to fall off prematurely. Commonly used as a pre-harvest treatment for soybeans and potatoes
DEFOLIATE	To strip off or to become stripped of leaves
DEGRADABILITY	The tendency of a chemical to decompose or break down into less toxic compounds or elements

DEPOSIT BUILDER	A material or mixture of additive materials which, in optimum quantity, improve the retention of a suspended formulation on a treated surface, reducing loss in the run-off of the liquid spray
DEPOSIT OR SPRAY RESIDUE	The amount of dry chemical deposited per unit area of plant, plant part, or other surface at any given time
DERMAL TOXICITY	The degree of toxicity or injuriousness of a test chemical when applied topically to the skin of a test animal
DESSICANT	A compound that promotes dehydration or removal of moisture from plant or other tissue which is intended for artificial acceleration of the drying of crop plants such as cotton leaves or potato vines
DETERGENT	A commonly used household cleaning agent that may be used as a wetting agent or spreader in pesticidal mixtures
DIATOMACEOUS EARTH	A whitish inert powder used as a pesticide formulation filler, or other inert ingredient in a pesticide formulation
DILUENT	A material, liquid or solid, serving to dilute the technical toxicant to field strength for adequate plant coverage, maximum effectiveness, and economy
DILUTION RATE	The amount of a diluent that must be added to a unit of pesticidal chemical to obtain the desired dosage
DISEASE	A condition in which any part of a living organism is abnormal: the condition of a plant that is being continuously affected by some factor that

	interferes with the normal activity of the plant's cells or organs
DISEASE COMPLEX	A plant disease that is caused by the interraction of two or more types of pathogens
DISEASE CYCLE	The sequence of events that occur between the time of infection and the final expression of disease symptoms
DISEASE ENDURANCE	*See* "Disease Tolerance"
DISEASE RANGE	The geographical distribution of a disease
DISEASE TOLERANCE	The ability of a host plant to survive and produce satisfactory yields at a level of infection that causes economic loss to other varieties or types of the same host plant
DISINFECTANT	A substance which destroys pathogenic fungi, bacteria, or other plant pathogens rendering them inactive
DISINFESTANT	A substance which destroys infesting organisms such as insects, mites, rats, weeds, and other multicellular organisms. In apposition to disinfectant
DISPERSING AGENT	A material that reduces the cohesive attraction between like particles
DISSEMINATION	The transport or spread of inoculum
DORMANT	A state of temporary inhibition of growth of seeds or other plant organs due to physiological causes
DOSE, DOSAGE	The quantity of a toxicant per unit of plant part, soil, or other surface. Frequently expressed as pounds per

	acre (kilograms per hectare), per thousand square feet, etc.
DOSE RATIO	Ratio between successively increasing dosages
DRIFT	The movement of a portion of the air-borne particles of a spray or dust application away from the intended site of application
DRY CONCENTRATE	A dry, relatively free-flowing powder containing the maximum possible amount of active ingredient. A wetting agent may be included so that the mixture is ready to be dispersed in water for spray application in which case it is referred to as "a dry wettable formulation"
ECONOMIC POISON	As defined under the Federal Insecticide, Fungicide and Rodenticide Act, an economic poison is "any substance or mixture of substances intended for preventing, destroying, repelling, or mitigating any insects, rodents, nematodes, fungi, or weeds, or any other forms of life declared to be pests." As so defined, economic poisons are now generally known as "Pesticides"
ECTOPARASITIC	A term used to characterize plant parasitic nematodes that feed on the root surface and normally do not enter the root tissue of the host. It is also descriptive of many parasitic seed plants that attach themselves to their hosts with a specialized absorbing root and form an aerial system of stems and leaves
ED_{50}	Effective Dose 50 is the preferred term for expressing the activity of a

fungicide and is synonymous with LD_{50}

EMULSIFIABLE CONCENTRATE — Produced by dissolving the toxicant and an emulsifying agent in an organic solvent. A solvent substantially insoluble in water is usually selected since water-miscible solvents have not in general proven to be feasible. The strength of the emulsifiable concentrate is usually expressed as pounds of toxicant per gallon of concentrate

EMULSIFIER — A surface-active substance which stabilizes (reduces the tendency to separate) a suspension of droplets of one liquid in another liquid which otherwise would not mix with the first liquid

EMULSION — A dispersion of fine particles of oily material in water (OW type) or less commonly as employed in pesticide usage, a dispersion of water droplets in oil (WO type). The one liquid does not dissolve in the other although a stable mixture results

ENATION — A disease symptom in which there is a mass of hypertrophied tissue on the surface of a leaf or stem

ENDEMIC — Natural to a country, geographical region, or area

ENDOPARASITE — A term used to characterize plant parasitic nematodes that invade, feed, and become established in the host root tissue, or attach themselves permanently to the root

ENDOSPORE — An asexual spore, developed within a cell, highly resistant, thick-walled, as

	in some bacteria. Capable of surviving exposure to high temperatures, or for long periods of dryness. The only plant parasitic bacteria to form endospores are in the genus Streptomyces
ENPHYTOTIC	A plant disease that occurs in low incidence more or less constantly in a given population
EPIDEMIC	A sudden destructive outbreak of a disease which normally does not cause a significant economic loss in a given area or region
EPINASTY	Increased growth on the upper surface of a plant organ or part (especially of leaves), causing it to bend downwards
EPIPHYTE	A plant which grows upon another plant, which it uses for mechanical support but not as a source of food
ERADICANT FUNGICIDE	An agent applied to a plant, plant part, or the environment to destroy established fungi within a given plant part or area of soil. Eradicant fungicides may function as disinfestants, disinfectants, or both. An eradicant fungicide destroys a fungus at its source after establishment within the host
ERADICATION	Complete elimination of a living organism from a given area or region
EXTENDER	An agent that extends the effective life of the applied treatment by protecting the residue from degrading environmental stresses
FAHRENHEIT (F)	A measure of temperature on a scale which marks the freezing point of

	water at 32°F and the boiling point of water at 212°F
FILLER	A diluent in powder form
FLASH POINT	The lowest temperature at which a combustible liquid will give off a flammable vapor which will burn momentarily
FULL COVERAGE	This term on a label signifies that the total volume of spray to be applied will thoroughly cover the crop being treated to a point of run-off
FUMIGANT	A substance or mixture of substances which produce a gas, vapor, fumes, or smoke intended to inactivate insects, fungi, bacteria, or rodents
FUNGICIDE	A chemical used on plants as a protectant against invasion by a plant pathogen
FUNGISTAT	A material that has the property of preventing growth or multiplication of fungi without necessarily destroying the organism after the latter has become established on or within the host
FUNGUS	A non-chlorophyll bearing allophyte. That is, all non-chlorophyll bearing plants of an order lower than mosses; for example, rusts, smuts, mildews, and other molds. They may cause destructive plant diseases. Plural "Fungi"
GALL	An overgrowth or swelling, more or less spherical, composed of unorganized cells

GENERAL RESISTANCE	A uniform resistance against all races of a pathogen. The level of resistance is usually only moderate and often influenced by temperature. It is usually inherited polygenically. Same as "horizontal resistance"
GENESTATIC	The ability of a chemical to prevent sporulation
GERMICIDE	A substance that kills microorganisms
GROWTH REGULATOR	An organic substance effective in minute quantities as a modifier of plant processes
HAZARD	The possibility that a pesticide or other chemical will cause damage or harm
HORIZONTAL RESISTANCE	*See* "General Resistance"
HOST	Any plant or animal attacked by, or harboring, a living parasite and from which the invader is obtaining its nourishment
HOST RANGE	The different species of plants that may be attacked by a specific pathogen
HYDROGEN ION CONCENTRATION	*See* pH
HYGROSCOPIC	The ability to absorb moisture from the atmosphere under normal conditions of temperature, pressure and humidity
HYPERPLASIA	The production of an abnormally large number of cells

HYPERTROPHY	An abnormal increase in the size of plant cells
HYPOPLASIA	The production of a subnormal number of cells
HYPOTROPHY	An abnormal decrease in the size of plant cells
IMMUNITY	The ability of plant to remain free from disease due to inherent characteristics of the plant
INCOMPATIBLE	Not capable of being mixed or used in combination with certain other materials
INDIGENOUS	Native to a particular area. Not introduced
INERT MATERIAL	Inactive ingredient. Any chemical, by itself non-toxic, included in a pesticide formulation to improve physical properties
INGREDIENTS	The active pesticide chemicals plus other inactive materials in a pesticide formulation or mixture that are listed on the manufacturer's label
INGREDIENT STATEMENT	A label shall state the name and percentage by weight of each active ingredient, together with the total percentage by weight of the inert ingredients. In the ingredient statement, all names shall be printed in type of the same size, and in such terms as to render them likely to be read under customary conditions of purchase
INHIBITOR	A bacterial inhibitor is a substance that will prevent the growth and mul-

	tiplication of bacteria and is applicable to other forms of life
INOCULUM	That portion of a pathogen capable of being disseminated to cause new infections
INVERT EMULSION	A dispersion of droplets of water in oil: a "WO type" of emulsion adapted to special uses in pest control. Also refers to viscous liquids adapted to application from the air to reduce drift hazards
IN VITRO	A term used to describe tests conducted under artificial conditions outside of the host
IN VIVO	A term used to describe tests conducted under natural conditions within the host
ION	An electrically charged particle, atom, molecule or radical in which the charge is due to the gain or loss of one or more electrons
LC_{50}	A symbol denoting median lethal concentration rather than dose as in the case of LD_{50}
LD_{50}	An abbreviation for median lethal dose (MLD). It refers to the amount of toxicant necessary to effect a 50% kill of the test organism. It is usually expressed in terms of weight of the chemical per unit of body weight
LESION	A localized area of diseased tissue
LIFE CYCLE	The progressive series of changes undergone by an organism. Or, lineal

	successions of organisms including sexual and asexual stages. With many fungi this is the stage, or series of stages, between one spore form and the production of the same spore form again
LIQUID CONCENTRATE	A pesticide formulated as a liquid. Emulsifiable concentrate (EC)
LOW-VOLATILE	A liquid or solid which does not evaporate quickly at normal room temperatures
LOW VOLUME SPRAY	This signifies that the total volume of spray to be applied is adequate to cover uniformly the crop being treated, but not to the point of run-off
METABOLITE	A compound derived in the case of a pesticide by chemical action upon the pesticide within a living organism. The action varies and the metabolite may be more toxic or less toxic than before. The same derivative in some cases may develop outside of a living organism
MG/KG	Refers to the amount of pesticide in milligrams per kilogram of body weight required to produce a desired effect
MICRON	A unit of measurement approximately equal to 1/25,000 in.
MILDEWPROOFER	A substance used to treat fabrics, paints, paper, etc., which prevents the growth of mildew for a relatively long time, or for the life of the material
MILLIGRAM (Mg)	A unit of measurement of mass. 1 oz = 28.349 milligrams

MISCIBLE LIQUIDS	Two or more liquids capable of being mixed and remaining mixed under normal conditions
MOLD	Any fungus with conspicuous profuse or woolly growth (mycelium or spore masses). Usually occurs in damp areas, on decaying organic matter, or on the surface of plant tissue
MONOCULTURE	The production of successive crops of the same plant species
MOSAIC	A term applied to variegated patterns of green and yellow in normally green leaves
MOTTLE	Leaf symptoms of irregular light and dark areas
NECROSIS	Death of plant cells usually resulting in affected tissue turning dark in color
NECROTIC	Dead
NEMATICIDE	A compound lethal to nematodes
NEMATODE	A member of a large group (phylum Nemathelminthes) also known as a threadworm, eelworm, etc. Some larger types are parasites on man and larger animals. Nematodes injurious to plants, sometimes referred to as roundworms, are slender, free-living, microscopic, wormlike organisms in the soil
NONIONIC	A surfactant that does not ionize is classed as "non-ionic," in contrast to anionic and cationic compounds
OBLIGATE PARASITE	A parasite that can develop only in

living tissue and cannot be cultured artificially

ORAL TOXICITY — The relative toxicity of a chemical when introduced through the mouth. Usually expressed as the Oral LD_{50}, or the dosage in milligrams per kilogram of body weight required to kill 50% of the test population when introduced orally in one dose

OSMOSIS — The transfer of materials that takes place through a semi-permeable membrane that separates two solutions, or between a solvent and a solution that tends to equalize their concentration

PANDEMIC — An epidemic that occurs over an extended geographical area or areas

PARASITE — An organism that lives in or on some living plant or animal (called a host) and obtains all or part of its nutrient from it

PARTS PER MILLION — A method of expressing concentration of chemicals or other proportions. One inch is one part per million in 16 miles. A teaspoonful of chemical applied to one acre would be 5 parts per million

PATHOGEN — An organism capable of inciting a disease. In general usage in plant protection this is usually a bacterium, fungus, virus, nematode, or parasitic plant

PENETRANT — Wetting agents that enhance the ability of a liquid to enter into pores of a substrate to penetrate the surface

PEST	Any organism which injures man, his property, or his environment, or which annoys or inconveniences man. Such organisms include fungi, weeds, insects, nematodes, birds and rodents
PESTICIDE	Any substance used for controlling, preventing, destroying, repelling, or mitigating undesirable effects of pests. Legally defined as "Economic Poisons" in the USDA Code Section 601. Includes fungicides, insecticides, nematicides, rodenticides, etc.
pH	The symbol for hydrogen ion concentration and is a measure of acidity and alkalinity within the range of pH 1.0–14.0. pH 7.0 is the midpoint where a compound is neither acid nor alkaline. Below pH 7.0 the substance is progressively more acid, and above it progressively more alkaline
PHYSICAL CONSTANTS	Numerical values such as molecular weight, melting point, boiling point, specific gravity, viscosity, etc.
PHYSICAL PROPERTIES	Examples are solubility, volatility, liquid or solid or gas, inflammability, etc.
PHYTOTOXICITY	The degree to which a compound or formulation is injurious to vegetation. It is specific for particular kinds or types of individual plants
PLANT GROWTH REGULATOR	A preparation which in minute amounts alters the behavior of ornamental or crop plants or the produce thereof, through physiological rather than physical action. It may act to accelerate or retard growth, prolong or break a dormant condition, pro-

	mote rooting, or other physiological changes
POISON (BIOCIDE)	A substance which, when absorbed by eating, drinking, breathing, or otherwise ingesting in relatively small quantities, causes illness, death, or retardation of growth or shortening of life.
PREDISPOSITION	The tendency of non-genetic conditions acting before infection, to increase the susceptibility of a plant to disease
PROTECTIVE FUNGICIDE	A chemical or physical agent that protects an uninfected plant or plant seed by killing or inhibiting the development of fungus spores or mycelium present on the surface of a plant
QUARANTINE	The restraints or restrictions upon any or all plants or other commodities suspected of being carriers of plant pests
RADIOACTIVE	Giving off atomic energy in the form of radiation such as alpha, gamma, or beta rays
RADIOISOTOPES	A broad class of elements capable of becoming radioactive and giving off atomic energy such as is detectable with a Geiger counter. Some radioisotopes occur naturally while others are produced artificially. The term is synonymous with radioactive elements and includes tracer elements
RANGE	Geographic distribution. Not to be confused with host range
RATE	The actual amount of pesticidal

	chemical (active ingredient) applied to a given surface area
RELATIVE HUMIDITY	The ratio of the quantity of water vapor present in the atmosphere to the quantity which would saturate at the existing temperature. The relative humidity, even at a uniform vapor level, changes with a change in temperature because the quantity of moisture vapor required to produce saturation will also change
RESIDUE	The quantity of a substance, especially of active pesticide, remaining in or on the surface of a crop, including livestock products
RETENTION	The ratio of the amount of fungicide, or other pesticide deposit at a given time in relation to the initial deposit. Initial retention is the ratio of initial deposit to the amount of pesticide actually applied to a given area of plant, plant part, or other surface
ROSETTE	A plant disease symptom characterized by shortened stems or branches resulting in a bunchy type of growth
RUSSET	A rough, brownish marking on leaves, fruits, or tubers which may be caused by insects, fungi, weather conditions, or chemicals
SAFENER	A substance which prevents objectionable changes when two or more substances, which otherwise would not be compatible, must be mixed
SANITATION	A term used for a number of cultural methods that reduce or may reduce production of pathogen inoculum, e.g.,

clean cultivation, burial of infected crop residues, removal or burning of infected plant parts, disinfection of tools and machinery, etc.

SCORCH — Burning of leaf tissue which frequently occurs along the veins or margins of the affected plant leaf

SEED TREATMENT — The use of a pesticide on seeds before they are planted. Used either for protection, decontamination, or disinfestation

SELECTIVITY — That characteristic whereby undesirable species are destroyed while others such as crop plants or beneficial insects are relatively unharmed. Generally applied to herbicides and to some fungicides

SELF-EMULSIFYING OIL — An oil that does not require the addition of an emulsifier in order to be suspended in water

SILVICIDE — A chemical used to destroy woody shrubs or trees

SLURRY — A thin, watery mixture such as liquid mud or cement. Fungicides are applied as slurries in special treating equipment for seed treatment to avoid the hazards sometimes encountered with dust formulations of chemical seed protectants

SOIL STERILANT — Any material that makes soil incapable of supporting plant growth or soil-borne microorganisms

SOIL INCORPORATION — Mechanical mixing of a pesticide with the soil

SOLUBILITY — The solubility of a gas, liquid, or solid

material in a liquid solvent is the maximum quantity of the substance which will dissolve in the solvent. Solubility varies with temperature, and hence, is usually expressed as solubility at a standard temperature such as 25°C

SOLVENT	A liquid such as water, kerosene, xylene, alcohol, or acetone used to dissolve a pesticidal chemical to form a solution
SOURCE OF INOCULUM	The place or object on or in which pathogen inoculum is produced
SPOT TREATMENT	A treatment directed at specific plants or areas rather than an overall application
SPRAY DRIFT	The movement of airborne spray particles from the intended area of application
SPREADER	An inert material added to a spray preparation to increase or improve the contact between the chemical and the surface to which it is applied
STICKER	A material added to a spray preparation to increase tenacity rather than to increase initial deposit
SUPPLEMENT	Adjuvant, auxiliary spray material. A material added to a pesticide to improve physical or chemical properties, e.g., spreaders, stickers, safeners, wetting agents
SURFACTANT	A material which facilitates or accentuates the surface-modifying properties of a pesticide formulation or preparation. Emulsifier, dispersing agent, spreader, wetting agent

SUSPENSION	A system constituted by a finely divided solid particle dispersed in a liquid, solid, or gas, e.g., a wettable powder suspended in water
SYMBIOSIS	The living together in close association of two dissimilar organisms usually resulting in a relationship useful to one or both
SYNDROME	The pattern of symptoms and signs of a disease
SYNERGISM	The phenomenon which results in a cooperative action between discrete agents such that the total effect is greater than the sum of the two components taken separately, e.g., sulfur and dichlone
SYNERGIST	A chemical which enhances the toxicity or effectiveness of another chemical to a degree greater than the sum total of the toxicity of the two components
SYNTHESIS	The bringing together of two or more materials to form a new compound
SYSTEMIC	A chemical that is capable of being absorbed by a plant system and transported throughout the plant
TECHNICAL MATERIAL	The pesticidal chemical as it is manufactured by a basic producer before formulation with other materials
TENACITY	Adherence. The resistance of a pesticide deposit to weathering as measured by retention
TENACITY INDEX	The ratio of the residual deposit over the initial deposit when subjected to prescribed washing techniques

THERMAL AEROSOL	A thermal aerosol is produced by blowing hot air past a hydraulic sprayer nozzle emitting an oil base spray. The air passes at high velocity through a combustion chamber heated to 800°−1200°F
TOLERANCE	The tolerance for a pesticide is defined by the amendment to the Federal Food and Drug Act as the minute trace permitted by federal authority to be present in or on raw agricultural commodities. According to the Food Additive Amendment it is the trace permitted in or on processed or semi-processed food or food products. Pesticides which are considered as highly poisonous are assigned a zero tolerance and other pesticides of low toxicity may be exempt from tolerance restrictions
TOXIC	Poisonous, injurious, or harmful to plants, animals, or humans
TOXICANT	An agent capable of exhibiting toxic properties
TOXICITY	The degree to which a substance is injurious or poisonous to plants, animals, or humans
TRACER ELEMENT	A radioactive element used in biological research to trace the path of a substance or to follow stages in a chemical reaction such as the pathway of metabolism of a nutrient or growth substance in plants and animals. Radioactive elements that have proven useful for tracer work are Carbon 14, Calcium 45, Cobalt 60, Strontium 90, and Phosphorus 32
VAPOR PRESSURE	The ability of a chemical to change

	to the vapor phase (evaporate). The higher the vapor pressure the faster the chemical evaporates
VIRICIDE	A substance capable of inactivating or suppressing the multiplication of a virus completely and permanently
VISCOSITY	The ability of a liquid to resist flowing. The higher the viscosity the slower the liquid will flow. The higher the temperature the lower the viscosity. The opposite of viscosity is "fluidity"
VOLATILE	A compound is said to be volatile when it changes from a liquid to a gas at ordinary temperatures on exposure to air
WETTABLE POWDER	A pesticidal chemical in dry form that may be mixed with water to form a suspension, e.g., 50% WP means a wettable powder containing 50% active ingredient

C. THE METRIC SYSTEM

(The following description of the Metric System is based on information supplied by the United States Department of Commerce, National Bureau of Standards, Washington, D.C.)

John Quincy Adams, sixth President of the United States, in his report to the Congress in 1821 stated:

> Weights and measures may be ranked among the necessaries of life to every individual of human society. They enter into the economical arrangements and daily concerns of every family. They are necessary to every occupation of human industry; to the distribution and security of every species of property; to every transaction of trade and commerce; to the labors of the husbandman; to the ingenuity of the artificer; to the studies of the philosopher; to the researches of the antiquarian, to the navigation of the mariner, and the marches of the soldier; to the exchange of peace, and all the operations of war; The knowledge of them, as in established use, is among the first elements of education, and is often learned by those who learn nothing else, not even to read and write. This knowledge is riveted in the memory by the habitual application of it to the employment of men throughout life
>
> John Quincy Adams
> Report to the Congress, 1821

Since the earliest of recorded times weights and measures have been important perquisites in human society. There is evidence from primitive societies to indicate that weights and measures were used for bartering for food and raw materials, for the construction of dwellings of an appropriate size and shape, and for other routine activities.

Man first used parts of the human body for units of measure, and seasons of the year for the measurement of time. Early Babylonian and Egyptian records and the Bible indicate that length was first measured by the forearm, hand, or finger, and time was measured by the periods of the sun, moon, and other heavenly bodies. When it was necessary to compare the capacities of containers such as gourds, or clay or metal vessels, they were filled with plant seeds which were then counted to measure the volumes. When means for weighing were invented, seeds and stones served as standards. For instance, the "carat," still used as a measure of value for gems, was derived from the carob seed.

In the U.S.A. the system of weights and measures was brought into the country from England by early colonists, and remained in general use until recently.

On December 23, 1975, President Ford signed into law the "Metric

Conversion Act," which established for the first time a national policy supporting metric measurement, and thus ended the dilemma that had prevailed throughout the history of the U.S.A. for more than 200 years. According to the Metric Conversion Act:

> It is therefore declared that the policy of the United States shall be to coordinate and plan the increasing use of the metric system in the United States and to establish a United States Metric Board
>
> (Metric Conversion Act of 1975)

George Washington, in his first message to Congress on January 8, 1790, raised the question of a more uniform system of weights and measures, and John Quincy Adams and Thomas Jefferson strongly supported the suggestion of George Washington. Both of the latter favored the French metric system.

Over 300 years ago the need for a single worldwide coordinated measurement system was recognized. In 1670 Gabriel Mouton, the Vicar of St. Paul in Lyons, France, suggested a comprehensive decimal measurement system. In 1790, during the French Revolution, the National Assembly of France asked the French Academy of Science to develop a standardized system of weights and measures. The commission appointed by the French Academy of Science did in fact create a system that was both simple and scientific. This ultimately became what is referred to internationally as "the metric system of weights and measures." In 1875 the International Treaty, known as the Metric Convention, set up well defined metric standards for length and mass, and established permanent procedures to recommend and adopt additional refinements in the metric system. Seventeen countries, including the U.S.A., signed the treaty. By 1900 a total of 35 nations, including the major countries of continental Europe and most of South American countries, had officially accepted the metric system.

Prior to the passage of the U.S. Metric Conversion Act of 1975 all of the major U.S. trading countries either had gone metric or were in the process of doing so. The only countries that had not adopted the metric system by 1975 were Brunei, a small sultanate in northwest Borneo, Burma, Liberia, and Yemen on the southwest Arabian peninsula. The British Commonwealth countries, with whom the U.S.A. shared the pound, shillings, pence, quarts, and yards system for so many years, were all committed to the metric system before the U.S. Metric Conversion Act. Canada has an unofficial goal of total commitment to the metric system by 1980.

The U.S.A. has been slow in adopting the metric system. However, some sectors of American life in the past several years have independently "gone metric" without federal action. At least 37 major U.S. corporations have announced their intentions of "metricising" their operations. The total annual sales of these corporations exceeds $130 billion and the list includes four of the nation's largest corporations. Some of the first metric-minded corporations in the U.S.A. included General Motors Corporation, International Business Machines, Chrysler Corporation, Caterpillar Tractor Co., Minnesota Mining and Manufacturing Co., the Ford Motor Company, International Harvester Company, and the Xerox Corporation. Although the "metricized" corporations generally include only manufacturers with overseas sales, recently retail corporations such as the J.C. Penney Co., Sears and Roebuck, and Montgomery Ward Company have climbed on the metric wagon.

Several government agencies have initiated metric activity by action ranging from appointment of metric committees to the use of metric publications and metric direction signs. These agencies include the Department of the Interior, the Federal Highway Administration, the Department of Health, Education and Welfare, and the National Weather Service. Very recently the State of Indiana Highway Department has started using metric highway signs, and it is not uncommon to see signs along the major highways reading "Indianapolis to Lafayette 103 Kilometers."

Switching to the metric system should be relatively easy for U.S. citizens since we are already making more frequent use of the metric system than is generally realized. For example, in international athletic competition television sportscasters regularly refer to swimming and field track events in meters rather than in yards or feet. Already we have 35 millimeter film, cigarettes that are 100 millimeters long, and even one millimeter longer than that. In the supermarkets on more and more packaged items the weights are expressed in grams. Most of the beer distributors are giving the contents of their product in liters as well as in fluid ounces. Canned goods are now labelled in both ounces and grams, and very shortly the contents of canned goods will be given in grams only.

When the Metric Conversion Act was approved by the U.S. Congress in 1975, 13 states had already erected some metric signs; Georgia, Maine, and California had already established metric commissions; Pennsylvania and Florida had cooperated with the Sun Oil Company to sell gasoline by the liter, North Dakota had adopted metric signs for its state parks, and 18 state legislatures had introduced bills to coordinate metric conversion.

For many years the metric system has been used in scientific publications, although not to the exclusion of the English system. Today, as problems in science become more complicated, efforts are being made to use the metric system more widely to simplify computation and education. The metric system for units of measurement is destined for universal adoption. The conversion tables that follow are intended to simplify acceptance of the metric system in the science of phytopathology. For further explanation of and information about the metric system, the following sources should be consulted:

Metric Studies Center. American Institute for Research. Box 1113, Palo Alto, California 94302

Public Affairs Office. George C. Marshall Space Flight Center, Alabama 35812

U.S. Metric Association. Sugarloaf Star Route, Boulder, Colorado 80302

BIBLIOGRAPHY

ANON. (ND). American National Metric Council. 1625 Massachussetts Avenue N.W. Washington, D.C. 20034.

BROWN, D. 1972. The metric system. The switch is coming. What it will mean. Crops Soils Aug–Sept. p. 13–14.

DE SIMONE, D.V. 1971. A metric America. A decision whose time has come. U.S. National Bureau of Standards Spec. Publ. 345. Washington, D.C.

HUNTOON, R.D. *et al.* 1970. International Standards (An interim report of the U.S. Metric Study). U.S. National Bureau of Standards Spec. Publ. 345–1.

NATIONAL BUREAU OF STANDARDS. 1975. A brief history of measurement systems with a chart of the modernized metric system. U.S. National Bureau of Standards Spec. Publ. 304 A.

TABLE A.1. METRIC CONVERSION FACTORS

Approximate Conversions to Metric Measure

Symbol	When you Know	Multiply by	To Find	Symbol
		LENGTH		
in.	inches	2.54	centimeters	cm
ft	feet	30	centimeters	cm
yd	yards	0.9	meters	m
mi	miles	1.6	kilometers	km
		AREA		
$in.^2$	square inches	6.5	sq centimeters	cm^2
ft^2	square feet	0.09	sq meters	m^2
yd^2	square yards	0.8	sq meters	m^2
mi^2	square miles	2.6	sq kilometers	km^2
	acres	0.4	hectares	ha
		MASS (Weight)		
oz	ounces	28	grams	g
lb	pounds	0.45	kilograms	kg
	short tons (2000 lb)	0.9	metric tons	t

TABLE A.1 ***(Continued)***

Symbol	When you Know	Multiply by	To Find	Symbol
		VOLUME		
tsp	teaspoons	5	milliliters	ml
tbsp	tablespoons	15	milliliters	ml
fl oz	fluid ounces	30	milliliters	ml
c	cups	0.24	liters	l
pt	pints	0.47	liters	l
qt	quarts	0.95	liters	l
gal.	gallons	3.8	liters	l
ft^3	cubic feet	0.03	cubic meters	m^3
yd^3	cubic yards	0.76	cubic meters	m^3
		TEMPERATURE (Exact)		
°F	Fahrenheit	5/9 after subtracting 32	Celsius	°C

For exact conversion factors refer to National Bureau of Standards. Misc. Publ. *286*. Units of weights and measures. Washington, D.C. 20402.

TABLE A.2. METRIC CONVERSION FACTORS

Approximate Conversions from Metric Measures

Symbol	When you Know	Multiply by	To Find	Symbol
		LENGTH		
mm	millimeters	0.04	inches	in.
cm	centimeters	0.4	inches	in.
m	meters	3.3	feet	ft
m	meters	1.1	yards	yd
km	kilometers	0.6	miles	mi
		AREA		
cm^2	sq centimeters	0.16	sq inches	in.^{2}
m^2	sq meters	1.2	sq yards	yd^2
km^2	sq kilometers	0.4	sq miles	mi^2
ha	hectares (10,000 m^2)	2.5	acres	
		MASS (Weight)		
g	grams	0.035	ounces	oz
kg	kilograms	2.2	pounds	lb
t	metric tons (1000 kg)	1.1	short tons	
		VOLUME		
ml	milliliters	0.03	fluid	fl oz
			ounces	pt
l	liters	2.1	pints	qt
l	liters	1.06	quarts	gal.
l	liters	0.26	gallons	ft^3
m^3	cubic meters	35	cubic feet	yd^3
m^3	cubic meters	1.3	cubic yards	

TABLE A.2 *(Continued)*

Symbol	When you Know	Multiply by	To Find	Symbol
		TEMPERATURE		
°C	Celsius	9/5 then add 32	Fahrenheit	°F

For exact conversion factors refer to National Bureau of Standards. Misc. Publ. *286*. Units of weights and measures. Washington, D.C. 20402.

D. CONVERSION FACTORS, TABLES AND FORMULAE

TABLE A.3. TEMPERATURE CONVERSION: CELSIUS TO FAHRENHEIT

To change temperature in degrees Celsius to degrees Fahrenheit:

Multiply degrees Celsius by 9/5 and add 32
Example: 30°C = 30 × 9/5 + 32 = 86°F

Degrees Celsius	Degrees Fahrenheit	Degrees Celsius	Degrees Fahrenheit
0	32.0	52	125.6
2	35.6	54	129.2
4	39.2	56	132.8
6	42.8	58	136.4
8	46.4	60	140.0
10	50.0	62	143.6
12	53.6	64	147.2
14	57.2	66	150.8
16	60.8	68	154.4
18	64.4	70	158.0
20	68.0	72	161.6
22	71.6	74	165.2
24	75.2	76	168.8
26	78.8	78	172.4
28	82.4	80	176.0
30	86.0	82	179.6
32	89.6	84	183.2
34	93.2	86	186.8
36	96.8	88	190.4
38	100.4	90	194.0
40	104.0	92	197.6
42	107.6	94	201.2
44	111.2	96	204.8
46	114.8	98	208.4
48	118.4	100	212.0
50	122.0		

TABLE A.4. TEMPERATURE CONVERSION: FAHRENHEIT TO CELSIUS

To change temperature in degrees Fahrenheit to degrees Celsius:

Subtract 32 from Fahrenheit figure and multiply by 5/9
Example: 86°F = 86 −32 × 5/9 = 50°C

Degrees Fahrenheit	Degrees Celsius	Degrees Fahrenheit	Degrees Celsius
0	−17.8	65	18.3
5	−15.0	70	21.1
10	−12.2	75	23.9
15	− 9.4	80	26.7
20	− 6.7	85	29.4
25	− 3.9	90	32.2
30	− 1.1	95	35.0
35	1.7	100	37.8
40	4.4	105	40.6
45	7.2	110	43.3
50	10.0	115	46.1
55	12.8	120	48.9
60	15.6	125	51.7

TABLE A.4 *(Continued)*

Degrees Fahrenheit	Degrees Celsius	Degrees Fahrenheit	Degrees Celsius
130	54.4	195	90.6
135	57.3	200	93.3
140	60.0	205	96.1
145	62.8	210	98.9
150	65.6	212	100.0
155	68.3		
160	71.1		
165	73.9		
170	76.7		
175	79.4		
180	82.2		
185	85.0		
190	87.8		

TABLE A.5. CONVERSION FACTORS FOR UNITS OF LENGTH

To Convert	To	Multiply by
centimeters	inches	0.3937
centimeters	meters	0.01
centimeters	millimeters	10.0
feet	centimeters	30.48
feet	inches	12.0
feet	yards	0.333
feet/minute	feet/second	0.0167
feet/minute	miles/hour	0.0014
feet	meters	0.3048
inches	centimeters	2.54
inches	feet	0.0833
inches	yards	0.0278
kilometers	feet	3281.00
kilometers	meters	1000.00
kilometers	miles	0.6214
kilometers	yards	1094.00
meters	centimeters	100.00
meters	feet	3.281
meters	inches	39.370
meters	kilometers	0.001
meters	yards	1.094
miles	feet	5280.00
miles	rods	320.00
miles	yards	1760.00
miles	kilometers	1.609
miles/hour	feet/minute	88.0
miles/hour	feet/second	1.467
miles/minute	feet/second	88.0
miles/minute	miles/hour	60.00
rods	feet	16.50
yards	feet	3.00
yards	inches	36.0
yards	meters	0.9144
yards	miles	0.00057

TABLE A.6. LENGTH CONVERSION: MILES TO KILOMETERS

To convert miles to kilometers multiply by 1.61.
To convert kilometers to miles multiply by 0.62.

+ Miles	0	1	2	3	4 Miles	5	6	7	8	9
					Kilometers					
0	—	1.61	3.22	4.83	6.44	8.05	9.66	11.27	12.88	14.48
10	16.09	17.70	19.31	20.92	22.53	24.14	25.75	27.36	28.97	30.58
20	32.19	33.80	35.41	37.02	38.62	40.23	41.84	43.45	45.06	46.67
30	48.28	49.89	51.50	53.11	54.72	56.33	57.94	59.55	61.16	62.76
40	64.37	65.98	67.59	69.20	70.81	72.42	74.03	75.64	77.25	78.86
50	80.47	82.08	83.69	85.30	86.90	88.51	90.12	91.73	93.34	94.95
60	96.56	98.17	99.78	101.39	103.00	104.61	106.22	107.83	109.44	111.04
70	112.65	114.26	115.87	117.48	119.09	120.70	122.31	123.92	125.53	127.14
80	128.75	130.36	131.97	133.58	135.19	136.79	138.40	140.01	141.62	143.23
90	144.84	146.45	148.06	149.67	151.28	152.89	154.50	156.11	157.72	159.33
100	160.93	162.54	164.15	165.76	167.37	168.98	170.59	172.20	173.81	175.42
Miles										

TABLE A.7. LENGTH CONVERSION: FEET TO METERS

To convert feet to meters multiply by 0.3048.
To convert meters to feet multiply by 3.2808.

+	0	1	2	3	4 Feet	5	6	7	8	9
					Meters					
0	—	0.31	0.61	0.91	1.22	1.52	1.83	2.13	2.44	2.74
10	3.05	3.35	3.66	3.96	4.27	4.57	4.88	5.18	5.49	5.79
20	6.10	6.40	6.71	7.01	7.40	7.62	7.93	8.23	8.53	8.84
30	9.14	9.45	9.75	10.06	10.36	10.67	10.97	11.28	11.58	11.89
40	12.19	12.50	12.80	13.11	13.41	13.72	14.02	14.33	14.63	14.94
50	15.24	15.54	15.85	16.15	16.46	16.76	17.07	17.37	17.68	17.98
60	18.29	18.59	18.90	19.20	19.51	19.81	20.12	20.42	20.73	21.03
70	21.34	21.64	21.95	22.25	22.56	22.86	23.16	23.47	23.77	24.08
80	24.38	24.69	24.99	25.30	25.60	25.91	26.21	26.52	26.82	27.13
90	27.43	27.74	28.04	28.35	28.65	28.96	29.26	29.57	29.87	30.18
100	30.48	30.78	31.09	31.39	31.70	32.00	32.31	32.61	32.92	33.22
Feet										

TABLE A.8. LENGTH CONVERSION: INCHES TO CENTIMETERS

To convert inches to centimeters multiply by 2.54.
To convert centimeters to inches multiply by 0.3937.

+	0	1	2	3	4	5	6	7	8	9
	Inches									
	Centimeters									
0	—	2.54	5.08	7.62	10.16	12.70	15.24	17.78	20.32	22.86
10	25.40	27.94	30.48	33.02	35.56	38.10	40.64	43.18	45.72	48.26
20	50.80	53.34	55.88	58.42	60.96	63.50	66.04	68.58	71.12	73.66
30	76.20	78.74	81.28	83.82	86.36	88.90	91.44	93.98	96.52	99.06
40	101.60	104.14	106.68	109.22	111.76	114.30	116.84	119.38	121.92	124.46
50	127.00	129.54	132.08	134.62	137.16	139.70	142.24	144.78	147.32	149.86
60	152.40	154.94	157.48	160.02	162.56	165.10	167.64	170.18	172.72	175.26
70	177.80	180.34	182.88	185.42	187.96	190.50	193.04	195.58	198.12	200.66
80	203.20	205.74	208.28	210.82	213.36	215.90	218.44	220.98	223.52	226.06
90	228.60	231.14	233.68	236.22	238.76	241.30	243.84	246.38	248.92	251.46
100	254.00	256.54	259.08	261.62	264.16	266.70	269.24	271.78	274.32	276.86
Inches										

TABLE A.9. CONVERSION FACTORS FOR UNITS OF MASS (WEIGHT)

To Convert	To	Multiply by
grains	grams	0.0648
grams	grains	15.432
grams	kilograms	0.001
grams	ounces	0.0353
grams/liter	parts/million	1000
kilograms	grams	1000
kilograms	pounds	2.205
ounces	grains	437.50
ounces	grams	28.35
ounces	pounds	0.0625
ounces (fluid)	cubic inches	1.805
ounces (fluid)	gallons	0.0078
ounces (fluid)	milliliters (ccs)	29.573
ounces (fluid)	pints	0.0625
ounces (fluid)	quarts	0.0313
parts/million	grains/U.S. gallon	0.0584
parts/million	grams/liter	0.001
parts/million	pounds/million gallons	8.345
pounds	grains	7000
pounds	grams	453.59
pounds	ounces	16.0
pounds	tons	0.0005
pounds of water	gallons	0.1198
tons	kilograms	907.185
tons	ounces	32000
tons	pounds	2000

TABLE A.10. MASS CONVERSION: POUNDS TO KILOGRAMS

To convert pounds to kilograms multiply by 0.454.
To convert kilograms to pounds multiply by 2.205.

+	0	1	2	3	4	5	6	7	8	9
	Pounds									
	Kilograms									
0	—	0.45	0.90	1.35	1.80	2.25	2.70	3.15	3.60	4.05
10	4.54	4.95	5.40	5.85	6.30	6.75	7.20	7.65	8.10	8.55
20	9.00	9.45	9.90	10.35	10.80	11.25	11.70	12.15	12.60	13.05
30	13.50	13.95	14.40	14.85	15.30	15.75	16.20	16.65	17.10	17.55
40	18.00	18.45	18.90	19.35	19.80	20.25	20.70	21.15	21.60	22.05
50	22.50	22.95	23.40	23.85	24.30	24.75	25.20	25.65	26.10	26.55
60	27.00	27.45	27.90	28.35	28.80	29.25	29.70	30.15	30.60	31.05
70	31.50	31.95	32.40	32.85	33.30	33.75	34.20	34.65	35.10	35.55
80	36.00	36.45	36.90	37.35	37.80	38.25	38.70	39.15	39.60	40.05
90	40.50	40.95	41.40	41.85	42.30	42.75	43.20	43.65	44.10	44.55
100	45.00	45.45	45.90	46.35	46.80	47.25	47.70	48.15	48.60	49.05
Pounds										

TABLE A.11. MASS CONVERSION: KILOGRAMS TO POUNDS

To convert kilograms to pounds multiply by 2.205.
To convert pounds to kilograms multiply by 0.454.

+	0	1	2	3	4	5	6	7	8	9
	Kilograms									
	Pounds									
0	—	2.21	4.41	6.62	8.82	11.03	13.23	15.44	17.64	19.85
10	22.05	24.26	26.46	28.67	30.87	33.08	35.28	37.49	39.69	41.90
20	44.10	46.31	48.51	50.72	52.92	55.13	57.33	59.54	61.74	63.95
30	66.15	68.36	70.56	72.77	74.97	77.18	79.38	81.59	83.79	86.00
40	88.2	90.41	92.61	94.82	97.02	99.23	101.43	103.64	105.84	108.05
50	110.25	112.46	114.66	116.87	119.07	121.28	123.48	125.69	127.89	130.10
60	132.28	134.51	136.71	138.92	141.12	143.33	145.53	147.74	149.94	152.15
70	154.35	156.55	158.76	160.97	163.17	165.38	167.58	169.79	171.99	174.20
80	176.40	178.60	180.80	183.02	185.22	187.43	189.63	191.84	194.04	196.25
90	198.45	200.66	202.86	205.06	207.27	209.48	211.68	213.89	216.09	218.30
100	220.50	222.71	224.91	227.12	229.32	231.53	233.73	235.94	238.14	240.35
Kilograms										

TABLE A.12. BUSHEL WEIGHTS OF COMMON COMMODITIES

Commodity	Weight of One Bushel (Average) Pounds	Kilograms
Alfalfa	60	27.0
Alsike Clover	60	27.0
Barley	48	21.6
Birdsfoot Trefoil	60	27.0
Buckwheat	14	6.3
Corn (shelled)	56	25.2
Field Beans	60	27.0
Flaxseed	56	25.2
Millet	50	22.5
Oats	32	14.4
Peas (field)	60	27.0
Potatoes	60	27.0
Redtop	14	6.3
Rice	45	20.25
Soybeans	60	27.0
Sudangrass	40	18.0
Timothy	45	20.3
Trefoil	60	27.0
Wheat	60	27.0

TABLE A.13. YIELD EQUIVALENTS

Kilograms/Hectare	Pounds/Acre	Bushels/acre 56 lb	Bushels/acre 60 lb
1	0.9	—	—
2	1.8	—	—
3	2.7	—	—
4	3.6	—	—
5	4.5	—	—
10	9.0	—	—
20	18.0	—	—
30	27.0	—	—
40	36.0	—	—
50	45.0	—	—
60	54.0	—	—
70	63.0	—	—
80	71.0	—	—
90	81.0	—	—
100	90.0	—	—
200	180.0	3.2	3.0
300	270.0	4.8	4.5
400	360.0	6.4	6.0
500	450.0	8.0	7.5
600	540.0	9.8	9.0
700	630.0	11.1	10.3
800	720.0	12.7	11.8
900	810.0	14.3	13.3
1,000	900.0	15.9	14.8
2,000	1800.0	32.1	30.0
	Tons/Acre		
3,000	1.25	44.6	41.6
4,000	1.75	62.5	58.3
5,000	2.25	80.4	75.0
6,000	2.75	98.2	91.7
7,000	3.0	107.1	100.0

TABLE A.13 *(Continued)*

Kilograms/Hectare	Pounds/Acre	Bushels/acre	
		56 lb	60 lb
8,000	3.5	125.0	116.7
9,000	4.0	142.9	133.3
10,000	4.5	160.7	150.0
11,000	5.0	178.6	166.7
12,000	5.5	196.4	183.3
13,000	6.00	214.3	200.0
14,000	6.5	232.1	216.6
15,000	7.0	250.0	233.3

1 ton = 2000 lb = 907.2 kg.

TABLE A.14. CONVERSION FACTORS FOR UNITS OF VOLUME

To Convert	To	Multiply by
fluid ounces	pints	0.06
fluid ounces	quarts	0.03
fluid ounces	gallons	0.0078
fluid ounces	milliliters	29.60
fluid ounces	liters	0.0295
pints	fluid ounces	16
pints	quarts	0.5
pints	gallons	0.125
pints	milliliters	473.17
pints	liters	0.4731
quarts	fluid ounces	32
quarts	pints	2
quarts	gallons	0.25
quarts	milliliters	946.33
quarts	liters	0.946
gallons	fluid ounces	128
gallons	pints	8
gallons	quarts	4
gallons	milliliters	3785.33
gallons	liters	3.785
milliliters	fluid ounces	0.0338
milliliters	pints	0.0021
milliliters	quarts	0.0010
milliliters	gallons	0.0002
milliliters	liters	0.0010
liters	fluid ounces	33.814
liters	pints	2.113
liters	quarts	1.056
liters	gallons	0.264
liters	milliliters	1000

TABLE A.15. VOLUME CONVERSION: GALLONS TO LITERS

To convert gallons to liters multiply by 3.786.
To convert liters to gallons multiply by 0.264.

+	0	1	2	3	4 Gallons	5	6	7	8	9
Gallons					Liters					
0	—	3.785	7.572	11.358	15.144	18.927	22.716	26.502	30.288	34.074
10	37.860	41.646	45.432	49.218	53.004	56.790	60.576	64.362	68.148	71.934
20	75.720	79.506	83.292	87.078	90.864	94.650	98.436	102.22	106.01	109.79
30	113.58	117.37	121.15	124.94	128.72	132.51	136.30	140.08	143.87	147.65
40	151.44	155.23	159.01	162.80	166.58	170.37	174.16	177.94	181.73	185.51
50	189.30	193.09	196.87	200.66	204.44	208.23	212.02	215.80	219.59	223.37
60	227.16	230.95	234.73	238.52	242.30	246.09	249.88	253.66	257.45	261.23
70	265.02	268.81	272.59	276.38	280.16	283.95	287.74	291.52	295.31	299.09
80	302.88	306.67	310.45	314.24	318.02	321.81	325.60	329.38	333.17	336.95
90	340.74	344.53	348.31	352.10	355.88	359.67	363.46	367.24	371.03	374.81
100	378.60	382.39	386.17	389.96	393.74	397.53	401.32	405.10	408.90	412.7

TABLE A.16. VOLUME CONVERSION: LITERS TO GALLONS

To convert liters to gallons multiply by 0.264.
To convert gallons to liters multiply by 3.786.

+	0	1	2	3	4	5	6	7	8	9
					Liters					
					Gallons					
0	—	0.264	0.53	0.79	1.06	1.32	1.58	1.85	2.11	2.38
10	2.64	2.90	3.17	3.43	3.70	3.96	4.22	4.49	4.75	5.02
20	5.28	5.54	5.81	6.07	6.34	6.60	6.86	7.13	7.39	7.66
30	7.92	8.18	8.45	8.71	8.98	9.24	9.50	9.77	10.03	10.30
40	10.56	10.82	11.09	11.35	11.62	11.88	12.14	12.41	12.67	12.94
50	13.20	13.46	13.73	13.99	14.26	14.52	14.78	15.05	15.31	15.58
60	15.84	16.10	16.37	16.63	16.90	17.16	17.42	17.69	17.95	18.22
70	18.48	18.74	19.01	19.27	19.54	19.80	20.06	20.33	20.59	20.86
80	21.12	21.38	21.65	21.91	22.18	22.44	22.70	22.97	23.23	23.50
90	23.76	24.02	24.29	24.55	24.82	25.08	25.34	25.61	25.87	26.14
100	26.40	26.66	26.93	27.19	27.46	27.72	27.98	28.25	28.51	28.78
Liters										

TABLE A.17. AMOUNT (GRAMS) OF CHEMICAL REQUIRED TO PREPARE DIFFERENT AMOUNTS OF SPRAY MIXTURE

Recommended Dose per 100 gal.[1]			Amount of Material Required to Prepare[2]						
		Dilution	100 gal.	50 gal.	25 gal.	20 gal.	10 gal.	5 gal.	1 gal.
oz	lb	to be Used	gm	gm	gm	gm	gm	gm	gm
1.0	0.06	1−13903	28.3	14.2	7.1	5.7	2.8	1.4	0.3
2.0	0.12	1−6952	56.7	28.3	14.2	11.3	5.7	2.8	0.6
3.0	0.18	1−4633	85.0	42.5	21.3	17.0	8.5	4.3	0.9
4.0	0.25	1−3337	113.4	56.7	28.4	22.7	11.3	5.7	1.1
6.7	0.41	1−2000	189.9	95.0	47.5	38.0	18.0	9.5	1.0
8.0	0.50	1−1668	226.8	113.4	56.7	45.4	22.7	11.3	2.3
8.3	0.52	1−1600	235.3	117.6	58.8	47.1	23.5	11.8	2.4
10.0	0.62	1−1345	283.5	141.7	70.9	56.7	28.3	14.2	2.8
11.1	0.69	1−1200	314.6	157.3	78.7	62.9	31.5	15.7	3.1
12.0	0.75	1−1112	340.2	170.1	85.0	68.0	34.0	17.0	3.4
13.4	0.83	1−1000	379.8	189.9	95.0	76.0	38.0	19.0	3.8
16.0	1.00	1−835	453.6	226.8	113.4	90.7	45.4	22.7	4.5
16.7	1.04	1−800	473.4	237.7	118.3	94.7	47.3	23.8	4.7
20.0	1.25	1−668	566.9	283.5	141.7	113.4	56.7	28.4	5.7
24.0	1.50	1−563	680.3	340.1	170.1	136.1	68.0	34.0	6.8
24.3	1.39	1−600	688.8	344.4	172.2	137.8	68.9	34.4	6.9
26.7	1.67	1−500	756.8	378.4	189.2	151.4	75.7	37.8	7.6
28.0	1.75	1−477	793.7	396.8	198.4	158.7	79.4	39.7	7.9
32.0	2.00	1−417	907.1	453.5	226.8	181.4	90.7	43.4	9.1
33.4	2.08	1−400	946.8	473.4	236.7	189.4	94.7	47.3	9.5
40.0	2.50	1−338	1133.8	566.9	283.5	226.8	113.4	56.7	11.3
44.0	2.75	1−303	1247.2	623.6	311.8	249.4	124.7	62.4	12.5
44.5	2.78	1−300	1261.4	630.7	315.3	252.2	126.1	63.1	12.6
48.0	3.00	1−278	1360.6	680.3	340.2	272.2	136.1	68.0	13.6
56.0	3.50	1−238	1587.4	793.7	396.8	317.7	158.7	79.4	15.9
64.0	4.00	1−209	1814.1	907.1	453.5	362.8	181.4	90.7	18.1
66.8	4.17	1−200	1893.5	946.8	473.4	378.8	189.4	94.7	18.9
72.0	4.50	1−185	2040.1	1020.0	510.0	408.0	204.0	102.0	20.4
80.0	5.00	1−167	2267.7	1133.8	566.9	453.5	226.8	113.4	22.7
96.0	6.00	1−139	2721.2	1360.6	680.3	544.2	272.1	136.1	27.2
112.0	7.00	1−119	3174.7	1587.4	793.7	634.9	317.5	158.7	31.7
128.0	8.00	1−104	3628.3	1814.1	907.1	725.7	362.8	181.4	36.3
133.5	8.35	1−100	3784.2	1892.1	946.0	756.8	378.4	189.2	37.8
144.0	9.00	1−93	4081.8	2040.9	1020.4	816.4	408.1	204.1	40.8
160.0	10.00	1−83	4535.3	2267.7	1133.8	907.1	453.5	226.8	45.4
171.0	16.69	1−50	4847.1	2423.6	1211.8	969.4	484.7	242.4	48.5

[1] 834,521 lb water per 100 gal.
[2] To convert to ounces or pounds divide by 28.35 or 453.59, respectively.

TABLE A.18. AMOUNT (TEASPOONFULS) OF POWDER REQUIRED TO PREPARE SMALL VOLUME OF SPRAY MIXTURE CONCENTRATIONS OF ½, 1, 1½ AND 2 LB/100 GAL.

Weight of Powder per Teaspoon	Number of Teaspoonfuls Required in Preparation of[1]											
	25 gal. of Spray				10 gal. of Spray				5 gal. of Spray			
(gm)	0.5	1.0	1.5	2.0	0.5	1.0	1.5	2.0	0.5	1.0	1.5	2.0
0.10	567	1134	1701	2268	227	454	680	907	113	227	340	454
0.20	284	567	850	1134	113	227	340	454	57	113	170	227
0.30	189	378	567	756	76	151	227	302	38	76	113	151
0.40	142	284	425	567	57	114	170	227	28	57	85	114
0.50	113	227	340	454	45	91	136	181	23	45	68	91
0.60	94	189	284	378	38	76	113	151	19	38	57	76
0.70	81	162	243	324	32	65	97	130	16	32	49	65

TABLE A.18 *(Continued)*

Weight of Powder per Teaspoon (gm)	Number of Teaspoonfuls Required in Preparation of[1] 25 gal. of Spray				10 gal. of Spray				5 gal. of Spray			
	0.5	1.0	1.5	2.0	0.5	1.0	1.5	2.0	0.5	1.0	1.5	2.0
0.80	71	142	213	284	28	57	85	113	14	28	42	57
0.90	63	126	189	252	25	50	76	101	13	25	38	50
1.00	57	113	170	827	23	45	68	91	11	23	34	45
1.10	52	103	155	206	21	41	62	82	10	21	31	41
1.20	47	94	142	189	19	38	57	76	9	19	28	38
1.30	44	87	131	174	17	35	52	70	9	17	26	35
1.40	40	81	122	162	16	32	49	65	8	16	24	32
1.50	38	76	113	151	15	30	45	60	8	15	23	30
1.60	35	71	106	142	14	28	42	57	7	14	21	28
1.70	33	67	100	133	13	27	40	53	7	13	20	27
1.80	32	63	94	126	13	25	38	50	6	13	19	25
1.90	30	60	90	119	12	24	36	48	6	12	18	24
2.00	28	57	85	113	11	23	34	45	6	11	17	23
2.10	27	54	81	108	11	22	32	43	5	11	16	22
2.20	26	52	77	103	10	21	31	41	5	10	15	21
2.30	25	49	74	99	10	20	30	39	5	10	15	20
2.40	24	47	71	94	9	19	28	38	5	9	14	19
2.50	23	45	68	91	9	18	27	36	5	9	14	18
2.60	22	44	65	87	9	17	26	35	4	9	13	17
2.70	21	42	63	84	8	17	25	34	4	8	13	17
2.80	20	40	61	81	8	16	24	32	4	8	12	16
2.90	20	40	59	78	8	16	23	31	4	8	12	16
3.00	19	38	57	76	8	15	23	30	4	8	11	12
3.20	18	35	53	71	7	14	21	28	4	7	11	14
3.40	17	33	50	67	7	13	20	27	3	7	10	13
3.60	16	32	47	63	6	13	19	25	3	6	9	13
3.80	15	30	45	60	6	12	18	24	3	6	9	12
4.00	14	28	43	57	6	11	17	23	3	6	8	11
5.00	11	23	34	45	5	9	14	18	2	5	7	9
6.00	9	19	28	38	4	8	11	15	2	4	6	8
7.00	8	16	24	32	3	6	10	13	2	3	5	6
8.00	7	14	21	28	3	6	8	11	1	3	4	6
9.00	6	13	19	25	3	5	8	10	1	3	4	5

[1] At recommended dosages of 0.5, 1.0, 1.5, 2.0 lb/100 gal.

TABLE A.19. RATES OF APPLICATION OF SPRAYS TO ROW CROPS

Distance Between Rows (ft)	Gallons per Acre	Quarts per 100 ft of Row	Feet of Row Covered by 1 gal.
	75	0.67	600
	100	1	400
	125	$1\frac{1}{6}$	341
1	150	$1\frac{1}{3}$	300
	175	$1\frac{2}{3}$	240
	200	$1\frac{5}{6}$	218
	75	2	194
	100	3	145
	125	3.5	116
3	150	4	97
	175	5	83
	200	5.5	73

TABLE A.20. GALLONS PER ACRE REQUIRED TO SPRAY ORCHARDS OF DIFFERENT PLANTING DISTANCES (SQUARE PLANTING)

Distance Between Trees	Gallons per Acre When per Tree Requirements are							
(feet)	2 gal.	5 gal.	7 gal.	9 gal.	10 gal.	12 gal.	15 gal.	20 gal.
18	268	670	938	1206	1340	1608	2010	2680
20	218	545	763	981	1090	1308	1635	2180
21	198	495	693	891	990	1188	1485	1980
22	180	450	630	810	900	1080	1350	1800
24	150	375	525	675	750	900	1125	1500
25	140	350	490	630	700	840	1050	1400
30	96	240	336	432	480	576	720	960
35	72	180	252	324	360	432	540	720
40	54	135	189	243	270	324	405	540
45	44	110	154	198	220	264	330	440
50	34	85	119	153	170	204	255	340

TABLE A.21. OPERATING CHART FOR TRACTOR BOOM SPRAYER

Teejet Tip No.	Pressure (lb)	Gallons per Acre 3 mph	4 mph	5 mph	7.5 mph
¼T 80015	25	11.7	8.8	7.1	4.7
or	30	12.9	9.7	7.7	5.2
¼TT 80015	35	13.9	10.4	8.3	5.6
100 mesh screen	40	14.9	11.1	8.9	6.0
¼T 8002	25	15.7	11.8	9.4	6.3
or	30	17.2	12.9	10.3	6.9
¼TT 8002	35	18.5	13.8	11.1	7.4
50 mesh screen	40	19.8	14.8	11.8	7.9
½T 8003	25	23.4	17.6	14.1	9.4
or	30	25.8	19.3	15.4	10.3
¼TT 8003	35	27.7	20.7	16.6	11.1
50 mesh screen	40	29.6	22.2	17.8	11.8

[1] From USDA Farmer's Bull. *2062*.

Spray Formula Calculation

A. The number of pounds of pesticide needed to mix a spray containing a given percentage of active ingredient may be calculated by using the following formula:

$$\frac{\text{Gallons of spray wanted} \times 8 \times \text{percentage of active ingredient wanted}}{\text{Percentage of pesticide needed}}$$

= Pounds of pesticide needed

e.g. How many pounds of 6% gamma BHC wettable powder are needed to make 50 gal. of spray containing 0.03% gamma BHC?

$$\frac{50 \times 8 \times 0.03}{6} = 2\text{ lb}$$

B. The percentage of active ingredient in a spray mixture may be calculated by using the following formula:

$$\frac{\text{Pounds of pesticide used} \times \text{percentage of active ingredient in pesticide}}{\text{Gallons of spray mixture} \times 8}$$

= Percentage of active ingredient in the spray

e.g. Two pounds of 6% gamma BHC wettable powder is mixed with 50 gal. of water. What percentage of BHC is in the spray?

$$\frac{2 \times 6}{50 \times 8} = 0.03\%$$

C. To calculate solution percentages use the formula given below:

Given: Two solutions of 8% and 15%
Wanted: A solution of 12%

To find how much of each solution to use, subtract the smaller figure from the larger figure *diagonally* as follows:

12 from 15 = 3
8 from 12 = 4

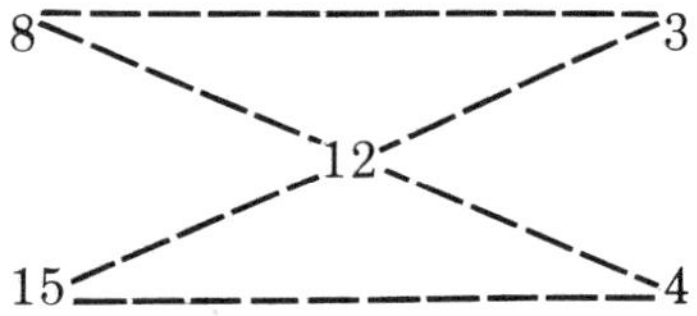

Now by reading *horizontally* you find that 3 parts of the 8% solution mixed with 4 parts of the 15% solution will give a 12% solution

SPRAY - APPLICATION FORMULA[1]

$$\text{Acres per minute} = \frac{2\text{ s.w.} \times \text{mph}}{1000}$$

$$\text{Gallons per minute} = \frac{2 \text{ s.w.} \times \text{mph} \times \text{gpa}}{1000}$$

s.w. = effective swath width in feet
mph = rate of travel in miles per hour
gpa = application rate in gallons per acre

TABLE A.22. SPRAY DRIFT AND DEPOSIT TABLE[1]

Drop Size (microns)	Visual Size Comparison	Time to Fall 10 ft (in Seconds)	Drift in 3 mph Wind (in Feet)	Drops per Square Inch if Applied at 1 gal. per Acre
5	Sea fog	4,050	18,000	9,216,00
33	Cloud	93	409	32,000
100	Mist	11	48	1,152
200	Drizzle	4.2	19	143
500	Light rain	1.59	7	9
1,000	Moderate rain	1.07	5	1
3,000	Cloudburst	0.87	4	1/23
6,000	Maximum drop size	0.83	3.5	1/182

1 micron = 1/1000 mm.
16 microns = 1/16 in.
[1] From USDA Farmer's Bull. *2062*.

FORMULA FOR CONCENTRATE SPRAYING

The term "concentrate spraying" has been introduced with the adoption of air-blast spraying equipment to define dosages of chemicals used and volume of spray material applied by this relatively recent spraying technique.

Concentrate spraying means the use of higher dosages of given chemicals in 100 gal. of water than is used in "conventional" or "dilute" spraying with hydraulic spray equipment. Concentrate spraying also involves the application of lower volumes of spray per tree to compensate for the higher dosages of toxicant used in the spray mixture. Concentrate spraying is used primarily for application of spray materials to fruit trees.

Dilutions designated as 2X, 3X, 6X, etc., are used to describe the degree of concentration in concentrate spraying. These dilutions simply mean that 2, 3 or 6 times the amount of chemical suggested for dilute spraying is used per 100 gal. of water, and that only one-half, one-third, or one-sixth the volume of spray material suggested for conventional spraying is applied per tree.

Before a sprayer can be calibrated for concentrate spraying the following information must be known:

1. Number of gallons of spray applied per tree by conventional or

dilute spraying
2. Average spread or ground area covered by the tree in feet
3. Speed of travel of the application equipment in feet per minute
4. The concentration of chemicals to be used; viz: 2X, 3X, or 6X, etc.

With this information available the following formula will give the required delivery in gallons per minute to be delivered from one side of the spray application equipment:

$$A = \frac{B \times C}{D \times E \times F}$$

where:

A = Spray delivery rate in gallons per minute for one side of the sprayer (figure to be determined)
B = The gallons of spray per tree used in dilute spraying
C = Speed of travel in miles per hour from which is calculated rate of travel in feet per minute
D = Always is 2 since only one-half of the tree is sprayed in passing
E = Concentration desired, viz., 2X, 3X, 4X, etc.
F = Average spread of trees to be sprayed in feet

Example:

Let us assume that:

B = 14 gal. of spray per tree required for dilute spraying
C = 4 mph (1 mph = 88 ft/min; 4 mph = 88 × 4 = 352 ft/min)
D = 2
E = A concentration of 4X is desired. E = 4
F = Average spread of trees being sprayed = 30 ft

then:

$$A = \frac{14 \times 352}{2 \times 4 \times 30} = \frac{4928}{240} = 20.5 \text{ gal./min}$$

Thus the delivery rate for one side of the sprayer should be 20.5 gal./min. For two-sided spraying each side of the sprayer should deliver 20.5 gal./min. This would make total of 41.0 gal./min using the series of nozzles on both sides of the sprayer.

Concentrate spraying delivers the same amount of chemical to the tree as would dilute spraying but results in savings in time and labor due to the need for less water in the spraying equipment.

CALIBRATION OF APPLICATION EQUIPMENT

Rate of application, granular and sprays, may vary with materials used. Application equipment must be calibrated for each material applied to obtain accurate delivery. Here are suggested steps to calibrating three types of applicators.

Boom Sprayer

1. Clean sprayer and replace all worn or defective parts.
2. Fill tank with water.
3. Adjust spray pressure and speed of tractor for nozzle size and output using manufacturer's directions.
4. Spray ¼ acre (10,890 sq ft). Distance of travel will vary with boom width. For example, a 22 ft boom must travel 495 ft to cover ¼ acre:

$$\frac{\text{¼ acre (10,890 sq ft)}}{\text{Boom width (22 ft)}} = \text{distance of travel (495 ft)}$$

5. Measure amount of water needed to refill the tank. This amount was applied to the ¼ acre; thus, four times this amount is the gallonage per acre.
6. Adjustment in gallonage may be made either by varying tractor speed or by changing nozzle size. Recalibrate after making an adjustment.
7. Calculate acres covered by tank of spray solution, and add required amount of herbicide for total area to be sprayed.

Band Sprayer

1. Clean sprayer and replace all worn or defective parts.
2. Fill tank with water.
3. Adjust spray pressure and speed of tractor for nozzle size and output using manufacturer's directions.
4. Spray ¼ acre (10,890 sq ft). Distance of travel will vary with number of rows on the planter and row width. For example, band spraying over 4 rows spaced 40 in. required 817 ft to cover ¼ acre:

$$\frac{\text{¼ acre (10,890 sq ft)}}{\text{Rows (4)} \times \text{row width (3.33 ft)}} = \frac{\text{10,890 sq ft}}{\text{13.33 ft}} = \text{distance of travel (817 ft)}$$

5. Measure amount of water needed to refill the tank. This amount was applied to the ¼ acre; thus, four times this amount is the

gallonage per acre.

6. Adjustment in gallonage may be made either by varying tractor speed or by changing nozzle size. Recalibrate after making an adjustment.
7. Calculate acres covered by tank of spray solution, and add required amount of herbicide for total *actual* area to be band treated.

Granular Band Applicator

1. Set applicator dial or dials to give desired delivery rate of granules, suggested for bank treatment, according to manufacturer's instructions.
2. Fill hoppers with granules to be used.
3. Travel across field at planting speed for the distance required to cover $\frac{1}{16}$ acre (2,722 sq ft) per row. Collect granules for *each* row in a bag, bucket or other container. For example, granular band application for 40-in. row requires 817 ft to cover $\frac{1}{16}$ acre:

$$\frac{\frac{1}{16}\text{ acre }(2{,}722\text{ sq ft})}{\text{Row width }(3.33\text{ ft})} = \text{distance to travel }(817\text{ ft})$$

4. Weigh granules from each row separately, and multiply by 16 to find delivery per acre for each row.
5. Adjust each setting, and recalibrate until the desired delivery rate is obtained.

Knapsack Sprayer

1. Lay out an area of 1 sq rod
   16.5 ft × 16.5 ft.
2. Record the time in minutes it takes to spray this area uniformly. Repeat this procedure for a total of three trials and average the three time intervals.
3. Catch the spray from the nozzle, or nozzles for the time interval (item 2) and measure it in pints.
4. Calculate the rate per acre,
   e.g., pints of spray caught (item 3) × 20 = gallons per acre.

E. BASIC PRODUCERS OF CHEMICALS FOR PLANT DISEASE CONTROL

Abbott Laboratories. Abbott Park, 1400 Sheridan Road. North Chicago, Illinois 60064

Aceto Chemical Co., Inc. 126-02 Northern Boulevard, Flushing, New York 11368

Allied Chemical Corporation. Agricultural Chemicals Division. 200 Race Street. Baltimore, Maryland 21230

Amchem Products, Inc. Brookside Avenue. Ambler, Pennsylvania 19002

American Cyanamid Co. Agricultural Chemicals Division. 50 Rockefeller Plaza. New York, New York 10020 or P.O. Box 400. Princeton, New Jersey 08540

American Hoechst Corporation. Agricultural Chemical Dept. P.O. Box 2500. Somerville, New Jersey 07110

The Ansul Company. 1 Stanton Street. Marinette, Wisconsin 54143

Ashland Chemical Company. Division of Ashland Oil Co. P.O. Box 2219. Columbus, Ohio 43216

Avitrol Corporation. P.O. Box 45141. Tulsa, Oklahoma 74145

BASF Aktiengesellschaft. Karl Bosch Strasse. 38. Ludwigshaven 67 (Rhein) West Germany

BASF Brasiliera. Praca D. Jose Gaspar 134. 10 Andar. Sao Paulo, Brazil

Bayer AG. 509 Leverkusen. Bayerwerk, West Germany

Bayer do Brasil. Rua Alexandria de Gusmao 606 (Santo amaro). Sao Paulo, Brazil

Biagra-Velsicol Produtos para Agricultura LTDA. Rua Dr Candido Espinheira 143. Sao Paulo, Brazil

The Boots Company Ltd. 1 Thane Road West. Nottingham 56111 England

Buckman Laboratories, Inc. k256 North McLean Boulevard. Memphis, Tennessee 38108

Calumet Division. Calumet & Hecla Inc. Calumet Avenue. Calumet, Michigan

CascoGard AB. P.O. Box 11033 S-10061. Stockholm, Sweden

Celamerck GMBH and Co. P.O. Box 202. 6507 Ingelheim, West Germany

Celanese Chemical Company. 1211 Avenue of the Americas. New York, New York 10036

Chemagro Agricultural Chemical Division, Mobay Chemical Corporation. P.O. Box 4913. Kansas City, Missouri 64120

Chempar Chemicals Co. 260 Madison Avenue. New York, New York 10016

Chevron Chemical Company. Ortho Division. 200 Bush Street. San Francisco, California 94104

Chipman Chemicals Ltd. 519 Parkdale Avenue North. P.O. Box 100. Postal Station 'C'. Hamilton, Ontario, Canada

Ciba-Geigy Corporation. Agricultural Division. P.O. Box 11422. Greensboro, North Carolina 27409

Ciba-Geigy Ltd. P.O. Box CH 400. Basle. 7. Switzerland

Ciba-Geigy Quimica SA. Avenida Santo Amaro 5.137 Sao Paulo, Brazil

Cities Service Company. P.O. Box 50360. Atlanta, Georgia 30302

W.A. Cleary Corporation. Vineland Chemical Company. 1049 Somerset Street. P.O. Box 10. Somerset, New Jersey 08873

Commercial Solvents Corporation. 245 Park Avenue. New York, New York 10017

Cooper McDougal & Robertson. San Ramon 765. Montevideo, Uruguay

Crystal Chemical Company. 1525 North Post Oak Road. Houston, Texas 77055

Delta Sociedade Commercial Ltda. Rua Florencio de Abreu. 126. CJ. 25. Sao Paulo, Brazil

Diamond Shamrock Chemical Company. Agricultural Chemical Division. 1100 Superior Avenue. Cleveland, Ohio 44114

DuPont do Brasil SA. Rua Consolacao 57. Sao Paulo, Brazil

E.I. DuPont de Nemours & Company. 1007 Market Street. Wilmington, Delaware 19898

Elanco Products Co. A Division of Eli Lilly & Co. Research and Development. P.O. Box 708. Greenfield, Indiana

Elanco Products Co. Division of Eli Lilly & Co. P.O. Box 1750. 740 South Alabama Street. Indianapolis, Indiana 46206

Farbwerks Hoechst AG. 6000 Frankfurt (Main) 71. Lyoner Strasse 10. Postfach 71 02 10. Frankfurt, West Germany

Fisons Corporation. Agricultural Chemical Division. 2 Preston Court. Bedford, Maine 01730

Food Machinery Corporation, Agricultural Chemical Division. 2000 Market Street, Philadelphia, Pennsylvania 19103 or 100 Niagara Street. Middleport, New York 14105

GAF Corporation. Chemicals Division. 140 West 51st Street. New York, New York 10020

Great Lakes Chemical Corporation. P.O. Box 2200. West Lafayette, Indiana 47906

Gulf Oil Chemicals Co. 9009 West 67th Street. Merriam, Kansas 66202

Harshaw Chemical Co. Division of Kewanee Oil Co. 1945 East 97th Street. Cleveland, Ohio 44106

Hercules Inc. 910 Market Street. Wilmington, Delaware 19899

Hoechst Aktiengesellschaft. Verhauf Landwirtschaft/Beratung. 6000 Frankfurt am Main. 71 Lyonerstrasse 30. Postf. 710210. West Germany

Hoechst do Brasil Quimica e Farmaceutica SA, Rua Braulio Gomes 36. Sao Paulo, Brazil

Hokko Chemical Industry Co. Ltd. Mitsui Bldg. No. 2 4-2 Nihonbashi. Hongoku-cho. Chuo-ku. Tokyo, Japan

Hokko do Brasil Industria Quimica Agropecuaria Ltda. Rua Apenos 970. Sao Paulo, Brazil

Hooker Chemical Company. Box 344. Niagara Falls, New York 14302

ICI United States Incorporated. Agricultural Chemicals Division. P.O. Box 208. Goldsboro, North Carolina 27530

Japan Agricultural Chemical Co. Ltd. Nihon Nohyaku Co. Ltd 2-5. 1-chome Nihonbashi Chuo-Ku. Tokyo, Japan

Kalo Laboratories Inc. 10236 Bunker Hill Ridge. Kansas City, Missouri 64114

Kemira Cy. P.O. Box 330. Helsinki 10, Finland

Kerr-McGee Chemical Corp. Kerr McGee Building. Oklahoma City, Oklahoma 73125

Kocide Chemical Corporation. 12701 Almeda Road. Houston, Texas 77045

Kumai Chemical Industry Co., Ltd. 6-2 Ohte-machi 2-chome Chiyoda-Ku. Tokyo, Japan

Los Angeles Chemical Company. 4545 Ardine Street. South Gate, California 90280

M & T Chemicals Inc. Subsidiary of American Can Co. Woodridge Road. Rahway, New Jersey 07065

Makhtesheim-Agan. Koor Chemicals. Crop Protection Division. P.O. Box 60. Beer-Sheva, Israel

Mallinckrodt Incorporated. P.O. Box 5439. St. Louis, Missouri 63147

McLaughlin Gormley King. 8810 10th Avenue North. Minneapolis, Minnesota 55427

Merck and Co. Chemical Division. 126 Lincoln Avenue. P.O. Box "M". Rahway, New Jersey 07065

Michigan Chemical Corp. 351 East Ohio Street. Chicago, Illinois 60611

Miller Chemical and Fertilizer Corporation. P.O. Box 333. Hanover, Pennsylvania 17331

Mineral Development and Research Corporation. 800 Briarcreek Road. Charlotte, North Carolina 28205

Mitsui Agricultural Chemicals. 2-5 Kasumigaseki-3-Chome Chuo-Ku. Tokyo, Japan

Mobil Chemical Co. Industrial Chemicals Division. P.O. Box 26683. Richmond, Virginia 23261

Monsanto Chemical Co. Industrial Chemicals Division. 800 North Lindbergh Boulevard. St. Louis, Missouri 63124

Montedison S.P.A. Agricultural Products Division. Pizza Della Republica 14/16. Milano, Italy

Nalco Chemical Co. 2901 Butterfield Road. Oak Brook, Illinois 60521

Niagara Chemical Division of Food Machinery Corporation. 100 Niagara Street. Middleport, New York 14105

Nihon Nohyaku Co. Ltd. Export Division. 2-5. 1 chome Ohtemachi Chiyodaku (Shin-Ohtemachi Building) Tokyo, Japan

Niklor Chemical Co., Inc. 2060 East 220th Street. Long Beach, California 90810

Nor-Am Agricultural Products Inc. 11710 Lake Avenue. Woodstock, Illinois 60098

Nortox Agroquimica SA. Rodovia Melo Peixoto. Kilometer 197 Aropongas, Brazil

Occidental Chemical Co. P.O. Box 1185. Houston, Texas 77001

Olin Corporation. P.O. Box 991. Little Rock, Arkansas 72203

Onyx Chemical Company. Division of Millmaster Onyx Corp. 190 Warren Street. Jersey City, New Jersey 07302

Pennwalt Corporation. Agricultural Chemicals Division. 1630 East Shaw Avenue. Suite 179. Fresno, California 93710

Charles Pfizer and Co., Inc. 630 Flushing Avenue. Brooklyn, New York 11206

Pfizer Quimica LTDA. Via Dutra Kilometer 391. Guarulhos, Brazil

Phelps Dodge Refining Corporation. 300 Park Avenue. New York, New York 10022

Pittsburgh Plate Glass Co. Corona Chemical Division. Technical Dept. Morristown, New Jersey 07960

Plant Protection Ltd. Subsidiary of ICI Ltd. Great Britain Fernhurst. Haslemere. Surrey, England

PPG Industries Incorporated. Chemical Division. One Gateway Center. Pittsburgh, Pennsylvania 15222

Rhodia Incorporated. Agricultural Division. 23 Belmont Drive. Somerset, New Jersey 08873

Rhodia Industrias Quimicas e Texteis SA. Avenida Antonio Cardoso 319. Santo Andre, Brazil

Roberts Chemicals Inc. Division of Security Chemicals Inc. P.O. Box 546. Nitro, West Virginia 25143

Rohm and Haas. Agricultural Division. North America. Independence Mall West. Philadelphia, Pennsylvania 19105

Sandoz Brasil SA. Rua Sao Francisco 500 (Santo Amaro) 82. Sao Paulo, Brazil

Sandoz Incorporated. Agrochemical Division, Postfach 4002. Basel 13, Switzerland

Sandos Incorporated. P.O. Box 1489. Homestead, Florida 33030

Schering AG, Berlin. Mullerstrasse 170-172 D-1 Berlin 65 Postfach 65.03.11 Berlin, Germany

Shell Chemical Co. 2401 Criw Canyon Road. San Ramon, California 94583

Sherwin-Williams Chemicals. P.O. Box 6520. Cleveland, Ohio 44101

Sobin Chemicals Incorporated. 52 Sobin Park. Boston, Massachusetts 02210

Stauffer Chemical Company. Agricultural Chemicals Division. Westport, Connecticut 06880

Stauffer Produtos Quimicas Ltda. Rua Mattheus Grou 604. Sao Paulo, Brazil

Thompson-Hayward Chemical Co. 5200 Speaker Road. Kansas City, Kansas 66110

Tower Chemical Co. P.O. Box 585. Clermont, Florida 32711

Transval Incorporated. P.O. Box 69. Marshall Road. Jacksonville, Arkansas 72076

Triangle Chemical Company. P.O. Box 4528. Macon, Georgia 31208

Tuco Division of the Upjohn Company. 7171 Portage Road. Kalamazoo, Michigan 49001

Union Carbide Corporation. Agricultural Systems. 270 Park Avenue. New York, New York 10017 or P.O. Box 1906. Salinas, California 93902

Uniroyal Incorporated. Emic Building. Spencer Street. Naugatuck, Connecticut 06770

U.S. Gypsum Company. 101 South Wacker Drive. Chicago, Illinois 60606

Velsicol Chemical Corporation. 341 East Ohio Street. Chicago, Illinois 60611

Vineland Chemical Co., Inc. West Wheat Road. P.O. Box 745. Vineland, New Jersey 08360

F. A USEFUL REFERENCE LIBRARY ON PLANT DISEASES

AGRIOS, G.A. 1969. Plant Pathology. Academic Press, New York.

ANDERSON, H.W. 1956. Diseases of Fruit Crops. McGraw-Hill Book Co., New York.

ANON. 1959. A Compendium of Plant Diseases. The Rohm & Haas Company, Philadelphia.

ANON. 1978. Professional Turf Manual. E.I. DuPont de Nemours & Company. Biochemicals Dept. Wilmington, Delaware.

BENARDE, M.A. 1970. Our Precarious Habitat. An Integrated Approach to Understanding Man's Effect on the Environment. W.W. Norton & Company, New York.

BUTLER, E.J., and S.G. JONES. 1949. Plant Pathology. Macmillan & Co., London.

CHESTER, K.S. 1947. The Nature and Prevention of Plant Diseases. 2nd edition. The Blackiston Co., Philadelphia.

CHRISTENSEN, C.M. 1951. The Molds and Man—an Introduction to the Fungi. University of Minnesota Press, Minneapolis.

CHUPP, C.A. and A. SHERF. 1960. Vegetable Diseases and Their Control. The Ronald Press Co., New York.

DICKSON, J.G. 1956. Diseases of Field Crops. McGraw-Hill Book Co., New York.

DODGE, B.O., and H.W. RICKETT. 1943. Diseases and Pests of Ornamental Plants. The Jacques-Catell Press, Lancaster, Pa.

FORSBERG, J.L. 1963. Diseases of Ornamental Plants. Univ. of Illinois, College of Agriculture Spec. Publ. *3*. Urbana.

GALLI, F. *et al.* 1968. Manual of Phytopathology. Biblioteca Agronomica Ceres. Sao Paulo, Brazil. (Portuguese)

HASSALL, K.A. 1969. World Crop Protection. Volume 2. Pesticides. CRC Press, Cleveland, Ohio.

HEALD, F.D. 1943. Introduction to Plant Pathology. 2nd edition. McGraw-Hill Book Co., New York.

HORSFALL, J.G., and E.B. COWLING. 1977. Plant Diseases; an Advanced Treatise. Academic Press, New York.

KENAGA, C.B. 1974. Principles of Plant Pathology. 2nd edition. Balt Publishers, Lafayette, Indiana.

LARGE, E.C. 1962. The Advance of the Fungi. Dover Publications, New York.

MAY, C. *et al.* 1953. Plant Diseases. The Yearbook of Agriculture. U.S. Dep. Agric. Washington, D.C.

MELHUS, I.E., and G.C. KENT. 1954. Elements of Plant Pathology. Macmillan, New York.

NORTHEN, H.T., and R.T. NORTHEN. 1954. The Secret of the Green Thumb. The Ronald Press Co., New York.

PIRONE, P.P., B.O. DODGE, and H.W. RICKETT. 1960. Diseases and Pests of Ornamental Plants. The Ronald Press Co., New York.

ROBERTS, D.A., and C.W. BOOTHROYD. 1972. Fundamentals of Plant Pathology. W.H. Freeman, San Francisco.

SHURTLEFF, M.A. 1966. How to Control Plant Diseases in Home and Garden. 2nd edition. Iowa State University Press, Ames.

STAKMAN, E.C., and J.G. HARRAR. 1957. Principles of Plant Pathology. The Ronald Press Co., New York.

STAPLEY, J.H., and P.C.H. GAYNER. 1969. World Crop Production. Volume 1. Pests and Diseases. CRC Press, Cleveland.

STEVENS, F.L. 1925. The Plant Disease Fungi. Macmillan, New York.

STEVENS, R.W. 1976. Pesticides in the Environment. Marcel Decker, New York.

STREET, R.B. 1969. Diseases of Cultivated Plants in the Southwest. The University of Arizona Press, Tucson.

TORGESON, D.C. 1967. Fungicides. An Advanced Treatise. Volume 1. Agricultural and Industrial Applications. Academic Press, New York.

WALKER, J.C. 1952. Diseases of Vegetable Crops. McGraw-Hill Book Co., New York

WALKER, J.C. 1969. Plant Pathology. 3rd edition. McGraw-Hill Book Co., New York.

WEBER, G.F. 1973. Bacterial and Fungal Diseases of Plants in the Tropics. University of Florida Press, Gainesville.

WESTCOTT, C. 1971. Plant Disease Handbook. 3rd edition. Van Nostrand Reinhold, New York.

Index

Other AVI Books

CITRUS SCIENCE AND TECHNOLOGY
Volume 2 *Nagy, Shaw and Veldhuis*

CORN: CULTURE, PROCESSING, PRODUCTS
Inglett

FOOD AND BEVERAGE MYCOLOGY
Beuchat

FUNDAMENTALS OF ENTOMOLOGY AND PLANT PATHOLOGY
2nd Edition *Pyenson*

FUNDAMENTALS OF FOOD MICROBIOLOGY
Fields

LABORATORY MANUAL FOR ENTOMOLOGY AND PLANT PATHOLOGY
Pyenson and Barkė

PHYSIOENGINEERING PRINCIPLES
Merva

PLANT PHYSIOLOGY IN RELATION TO HORTICULTURE
American Edition *Bleasdale*

POTATO PROCESSING
3rd Edition *Talburt and Smith*

POTATOES: PRODUCTION, STORING, PROCESSING
Smith

SMALL FRUIT CULTURE
5th Edition *Shoemaker*

SORGHUM PRODUCTION AND UTILIZATION
Wall and Ross

SWEET POTATOES: PRODUCTION, PROCESSING, MARKETING
Edmond and Ammerman

TOMATO PRODUCTION, PROCESSING AND QUALITY EVALUATION
Gould

TREE NUTS: PRODUCTION, PROCESSING, PRODUCTS
2nd Edition *Woodroof*

VEGETABLE GROWING HANDBOOK
Splittstoesser